Systemic Management

Sustainable Human Interactions with Ecosystems and the Biosphere

Charles W. Fowler

National Marine Mammal Laboratory
Seattle, Washington

The views and opinions expressed herein are strictly those of the author and do not reflect policy of the National Marine Fisheries Service or the Department of Commerce.

OXFORD
UNIVERSITY PRESS

OXFORD

UNIVERSITY PRESS

Great Clarendon Street, Oxford OX2 6DP

Oxford University Press is a department of the University of Oxford.
It furthers the University's objective of excellence in research, scholarship,
and education by publishing worldwide in

Oxford New York

Auckland Cape Town Dar es Salaam Hong Kong Karachi
Kuala Lumpur Madrid Melbourne Mexico City Nairobi
New Delhi Shanghai Taipei Toronto

With offices in

Argentina Austria Brazil Chile Czech Republic France Greece
Guatemala Hungary Italy Japan Poland Portugal Singapore
South Korea Switzerland Thailand Turkey Ukraine Vietnam

Oxford is a registered trade mark of Oxford University Press
in the UK and in certain other countries

Published in the United States
by Oxford University Press Inc., New York

© Oxford University Press 2009

British Library Cataloguing in Publication Data

Data available
10 06614364
Library of Congress Cataloging in Publication Data

Data available

Typeset by Newgen Imaging Systems (P) Ltd., Chennai, India
Printed in Great Britain
on acid-free paper by
CPI Antony Rowe, Chippenham, Wiltshire

ISBN 978–0–19–954096–9 (hbk); 978–0–19–956759–1 (pbk)

10 9 8 7 6 5 4 3 2 1

About the cover

The art on the cover of this book was painted by Katherine Zecca, former graphic designer and scientific illustrator at the Alaska Fisheries Science Center (NOAA, NMFS) in Seattle, Washington, and now a freelance artist in Vermont (http://www.katherinezecca.com/). Typical of the excellence of her work, this piece shows some of the patterns and processes characteristic of ecosystems. Primary producers are found on the left with species from increasingly higher trophic categories toward the right. Species with small bodies are found at the bottom and the largest species are found at the top. Partially hidden in various locations are species that represent processes and characteristics of species and ecosystems. These include extinction (a fish), mobility (a bat), and symbiotic interactions (a humming bird), and emerging diseases (an AIDS virus).

Original requests for this art expressed the hope that a hidden representation of humans, to symbolize their influence on, and place in, nature, could be included. The difficulty of the task was rendered moot when readers pointed out that the ear of the elephant contains three different hidden images of human faces. These were totally unintentional, making their presence even more symbolic of the message of this book.

Dedication

In our every deliberation we must consider the impact of our decisions
on the next seven generations.

Great law of the Hau de no Saunee*

This book is dedicated to (and in memory of) my father, Ervon W.
Fowler, for his lessons in thought and to (and in memory of) my
mother, Merna, who taught me the love of reality. It is also dedicated to
Dr Gordon Orians whose incomparable insight regarding the natural
world is reflected in my thinking, to my wonderful wife, Jean, for her
enduring support and love, and finally, and most importantly, to the
next seven generations of all species.

* Versions of the "law of the seventh generation" similar to that quoted here are frequently found in expressions of
responsibility for our future. Some of its historical origins are documented conceptually in Fenton (1968), Parker (1916),
Scott (1912), and Wallace (1946). As indicated by Hodgson (1989): "The law of the seven generations is found in the Bible
and in the teachings of many cultures around the world. Among the First Nations, the Iroquois maintain it is a gift from
our ancestors, whose overwhelming wisdom is inherent in its culture, political and philosophical definition. As Chiefs,
as Leaders, we are to consider carefully, the effects of our decisions on the seven generations yet to come."

Preface

We can never solve our significant problems from the same level of thinking
we were at when we created the problem.

—Albert Einstein

Humans are now reawakening to an awareness of themselves as part of the vastly complex web of life. In part, we are realizing the enormous influence our species has and the extent to which we—in our hubris—have disrupted this web, risking extinction of our own species along with countless others. Can humanity find for itself a sustainable niche within the web of life? What serves to define that niche? These are the ultimate questions addressed by this book.

Such questions were hardly so clear when this book was first conceived. In the mid-1980s, the United States' National Marine Mammal Laboratory was reorganized, partly with the intent of converting from a species-oriented structure to a discipline-oriented structure. This change was related to decisions within the National Marine Fisheries Service, along with other government and nongovernmental organizations, to take seriously management as it would involve ecosystems. I became responsible for contributing to a scientific basis for management at the ecosystem level.

I soon learned, however, that we only make a small step in the right direction when ecosystems are added to what we consider. It became clear that extending conventional concepts of management to ecosystems is causing more problems than are being solved. Full-scale change is needed, and a major part involves the way we think—our beliefs. One of the central themes of this book is that of identifying the information that most realistically provides guidance—a very select part of the information revealed by science.

I came to realize that we cannot control the nonhuman; control is limited primarily to human action. Reining in our incredible influential power serves as a form of management. The option of minimizing abnormality, to achieve health (at all levels) becomes real. We can limit our interactions with other species, with ecosystems, and with individuals (Christensen *et al.* 1996, Fowler 2005, Fowler and Hobbs 2002, Mangel *et al.* 1996). Thus, the option of managing our use of ecosystems to mimic other species emerged in my thinking as an option that has its parallels at all levels of biological organization. We, as individuals, and as a species, can manage our use of, interactions with, and influence on, other individuals, other species, ecosystems, and the biosphere. Our species is just one part of the picture. This reality requires us to face questions about the constraints and actions our species should require of itself. We need to recognize that we are part of various systems that include other species faced with the same reality and complexity that we face. Fitting in means being a normal part.

Early in my work on this book, it seemed to me that the problems that confront us could be dealt with by further consideration, acceptance, and integration of the concept of extinction—especially selective extinction and speciation. However, it soon became clear that this was not going to suffice for management either. Like other fields of science, it provides critical understanding, but could never guide decision-making by itself. Whole systems thinking also greatly influenced my own thinking. All systems are finite while complexity and reality are not; there had to be an approach that was not confined to any one system or level in the hierarchical structure of life. All had to be taken into account.

A different perspective began to take form. The form of management that emerged was systemic management to minimize human abnormality. Realizing the potential was one of the most gratifying experiences of my career. Systemic management builds on the variety of ways we understand reality. With a major reassignment of the role of stakeholders, it conforms to the requirements of management as laid out in the literature.

In my attempt to digest the literature on management at the ecosystem level, I found frequent mention of the need for "unifying" perspectives; holism was needed. As I proceeded, the needs and the potential seemed to match. What we study are the parts of reality in its unified form: holism is found in reality; informative natural patterns emerge in this reality, integrating through emergence.

This book, then, is intended to open the door to a conversion in thinking, similar to what I experienced in the process of writing it. I see a need for the recognition and application of information more directly related to reality. All of the things that science studies are combined in nature—the reality we want to be part of in finding sustainability. That combination is what all sciences would account for if their disciplines could be united. My hope is that humans can find sustainability in this context.

I expect controversy—controversy stems naturally from change. Most controversial, perhaps, will be the conclusion that the role of values and opinions needs to be replaced. They take on an entirely different role in systemic management: motivation for asking clear management questions.

I identify problems in the conventional use of science. This is not to be misinterpreted; research and scientists are not to be branded as wrong or ill-intended. Rather than demonize science, my message concerns the identification and use of the products of science that best meet our needs for guidance.

I wrote this book for the benefit of the biota of the earth, focusing on action we humans can take, primarily as a species. As parents discover, we are not born with user guides: one title suggested for the book was "How to be a successful species." This book is part of my contribution to human success through my own individual action. If the messages and implications are troublesome, we must be mindful that benefits at the species level often require sacrifice at the individual level, sacrifices that become more extreme when ecosystems and the biosphere are included. In the spectrum of time scales, short-term benefits involve long-term costs and, conversely, short-term efforts involve long-term gains. This book is my best effort toward thinking of the welfare of future generations of the human species, and all those species that accompany and support us in life. I hope it will lead to substantive progress.

Acknowledgments

I could never have produced this book alone. I benefitted beyond measure from the support, challenge, and insight, from the complements and criticisms, and from the advice and difference of opinion from others—all immeasurable contributions to an improved product.

My phenomenal wife, Jean, supported me through the entire process, tolerating extensive hours spent working at home, and enduring diversions that true passions entail. There can be no adequate expression of my thanks to her!

Howard Braham, R.V. Miller, Doug DeMaster, Sue Moore and John Bengtson, Peter Boveng, Phil Clapham, Bob DeLong, Tom Gelatt, Tom Loughlin, and Paul Wade were supervisors and fellow program leaders at the National Marine Mammal Laboratory during my work on this book. These understanding people sacrificed time on my behalf and provided much needed support and encouragement.

If there are others who deserve co-authorship in this work, they would clearly include Jason Baker, Larry Hobbs and Dorothy Craig. Jason provided invaluable help in editing various versions of the book, wrestling with consistency, and endless discussions regarding the weaving of concepts into an intelligible whole. Larry has been absolutely invaluable and a constant source of support and wisdom, ideas and thinking throughout the process. The importance of his contributions is beyond measure. Dorothy brought both a passion and understanding of my message as well as a sharp editorial mind. I have learned as much in working with these three as I have in any other aspect of this project. In a similarly personal way, I want to thank Shannon McCluskey and Cynthia Hedlund for reminding me of myself—easily lost in the mission and passions behind this book.

I thank both Alec MacCall, and Andrea Belgrano for their encouragement, interest, and suggestions for getting this book published. Alec provided more reviews of this book than any other person; his contributions are deeply appreciated. Andrea's suggestion that Oxford University Press should be the publisher was pivotal in getting to this point.

I thank my colleagues at the Alaska Fisheries Science Center in Seattle, especially those who contributed to the final review of the manuscript: Phil Clapham, Bob DeLong, Sharon Melin, Susan Picquelle, Jay VerHoef, and Paul Wade.

Over the years, the immense task of getting to this point has involved valuable reviews (including those of related foundational papers), many discussions, criticisms, and challenges, moral support, altered schedules, art and graphs, suggested references, information, and data extracted from a wide variety of sources. Numerous interns and students have been invaluable in searching literature, finding data, and writing reports. In seeking publication, several editors have provided invaluable comments and suggestions as did reviewers from whom they sought advice. I requested reviews from a variety of fields of science, from various friends and from many colleagues—all of them were invaluable in their help. Earlier drafts of this book were used in university-level courses at several universities, and I appreciate the criticisms, comments, and suggestions that were generated. Collectively they are too numerous to list, but I extend my heartfelt appreciation to each and every one.

In addition to those listed above, Jim Anderson, Eric Charnov, Jerry Cufley, Alec MacCall, Jim MacMahon, Marty Nelson, Mike Perez, Tim Ragen, Tim Smith, and Paula White have been a continuing source of inspiration and our many thought-provoking conversations have been invaluable to me. Special thanks go to Bob Francis and Marc Mangel who coordinated two of the upper-division university courses that used the manuscript for this book as a text.

The kindness of Jim Brown, Robert Furness, Jane Masterson, Melanie Moses, Mark Pagel, Robert Peters, Robin Phillips, and Douglas Kelt for their efforts and time in providing data and graphic material is gratefully acknowledged.

Various communities of people have contributed to foundations upon which this book rests. These include those who have developed (and continue to develop) the various fields of science—especially those focusing on selective extinction/speciation (see Okasha 2006), ecology, and macroecology. The time, money, and personal expense behind revealing pattern-based information is immense and cannot go without being acknowledged. Another group includes those who have spent considerable time and effort in developing the concepts of management (e.g., see the references of Appendices 4.1 and 4.3 and workers to whom these references lead). This extends to those who have delved into epistemology and the role of human thought in psychology, values, decision making, and control. Here, people like Gregory Bateson count heavily in my mind. I am grateful to all the people who have contributed to thinking about management. There is a clear precedence to the message that we can learn from nature in the works of people like Aldo Leopold, John Muir and Wendell Berry—an innumerable group of thinkers (including many from the world's ancient wisdom traditions) whose contributions I hope I have at least partially translated to a scientifically based way of proceeding. To those whom I have overlooked in the many hours of searching the relevant literature, I apologize. I thank them all!

The editorial help of Dorothy Craig, Gary Duker and Jim Lee was invaluable. My sincere gratitude to the editors and staff at Oxford University Press, especially Ian Sherman, Helen Eaton, Carol Bestley, Caroline Broughton, and Kathy Lahav—tireless dedicated people.

This book is what it is, in part, as a result of everyone's contribution. It remains a first step, however, and only the future can reveal how blind I remain in my human view of that future and the reality to be faced in being part of it.

Contents

CHAPTER 1

Systemic management—what and why?

> When we try to pick out anything by itself, we find it hitched to everything else in the Universe.
>
> —John Muir

We are faced with an uncertain future. Species are disappearing. Deforestation, agriculture, and fishing have changed ecosystems. Our polluted Earth is warming, the pH of our oceans is declining, and introduced species are altering their habitat. Scientists discover more problems every day. We may have evolved to become an extinction-prone species. Can humans manage other species, ecosystems, or the Earth in response to this information? No. Because of the complexity and interconnected nature of reality, management that seeks to control, dominate, or design such systems is doomed to ultimate failure. Managing the nonhuman usually causes more problems than it solves.

Within limits, what we *can* manage is *ourselves*—in essence, reinventing ourselves as a species (Berry 1999). We *can* manage our interactions, influence, and relationships with each other and with the nonhuman. Managing to find an appropriate fit or place in our universe ("systemic management", Fowler 2003) requires self-control. This book explores systemic management, its applications, and foundation. Such management has a broad set of objectives, with the primary goal being sustainability for both humans and the nonhuman.[1] This means striving toward a future in which ecosystems and the biosphere can support a diversity of life, including humans. Ecosystems and other living systems respond to human influence and will respond to a sustainable human presence. One of the conclusions of this book is that immense change will be required; current

efforts are exposed as largely artificial, superficial, and fallacious.

This book has two primary objectives:

1. *To propose systemic management as a replacement for conventional forms of management.* Systemic management emerges from the tenets developed for management in the last several decades. These include our understanding of ourselves (e.g., as individuals, societies, and a species), our thinking and our belief systems. Systemic management is reality-based management with a variety of components that include ecosystem-based management. In its application, systemic management fully accounts for the complexity of nature and avoids the errors of conventional management.

2. *To describe an empirical objective methodology that provides guidance for achieving sustainable relationships between humans and the nonhuman.* Carefully using natural patterns (Belgrano and Fowler 2008, Fowler 2003, Fowler and Hobbs 2002) to guide human endeavor achieves objectivity not found in conventional management. The selection of patterns that will meet our needs requires that we ask our management questions with extreme care. Science then reveals patterns that match these questions and provide answers that prevent the misguidance of conventional management. Such science best meets the needs of management.

Although systemic management applies generally, the examples in this book focus primarily on sustainability among species. Empirical (e.g., macroecological) patterns reveal what it means to be sustainable as a species. Species reflect the numerous and often very powerful forces of nature. Our species has the choice of living within the limits of nature or facing the consequences of being anomalous—one of which is an untimely extinction.

Defining what it takes and then being a successful species are both complex, nontrivial matters. Complicated relationships with our environment include countless interactions with other species, ecosystems, and the biosphere (e.g., our use of water, consumption of food, and production of CO_2). Management involves social, religious, ethnic, political, and economic elements, exemplified by decision making, laws, policy, and action. Systemic management involves actions and decisions by individuals, institutions, and society—all carried out consistently (Hobbs and Fowler 2008). Being in accord with the laws of nature by mimicking empirical examples of sustainability results in consistency and objectivity. Because empirical examples reflect the complexity of their emergence, systemic management involves context and the complexity of the biological systems of which we are a part. It involves time scales we ordinarily ignore. It involves thinking about our thinking so as to embrace a basis for management that replaces thinking. It means abandoning the artificial and adopting the real. It means embracing the holistic of reality (Appendix 1.1).

This chapter summarizes the underlying concepts of the book including:

• An introduction to systemic management as it emerges from published tenets of management, derived in part from the recognized need for ecosystem applications (Fowler 2003).
• A review of the motivation to reject current management and accept a preferable, more objective, alternative.
• An introduction to some of the steps involved in systemic management as applied to the eastern Bering Sea, an example of the ecosystem component of systemic management that will be developed at the end of each subsequent chapter.
• A summary and preview of the progression of concepts found in the chapters ahead.

1.1 Tenets of management

Management includes our use of natural resources, our interaction with other species, ecosystems, and the biosphere. Great importance was placed on defining the principles of such management in the late 20th century (Rockford *et al.* 2008). Much of this effort was motivated by the need for management that applies to ecosystems (e.g., Christensen *et al.* 1996, McCormick 1999). The resulting work has been synthesized on numerous occasions (e.g., Arkema *et al.* 2006, Fowler 2003, Francis *et al.* 2007, Lackey 1998). With one primary exception, these tenets are more completely embodied in systemic management than any other approach. That exception involves the role of stakeholders (Fig. 1.1); it is a major exception. In conventional management, stakeholders are involved in setting objectives and making decisions. In systemic management, the role of stakeholders is confined to asking clear management and science questions, followed by carrying out management with objectives inherent to the information produced by science. Objectivity is achieved in this shift—systemic management is pattern-based rather than opinion-based management (Belgrano and Fowler 2008). The science behind systemic management provides information that needs no conversion; the translation (Brosnan and Groom 2006) inherent to current management becomes unnecessary in answering management questions systemically.

Nine primary tenets of management are introduced below. Following each tenet is a very brief description of how systemic management simultaneously and consistently adheres to the principles involved.

Management Tenet 1: Management must be based on an understanding of humans as part of complex biological systems (Christensen *et al.* 1996, Mangel *et al.* 1996, NRC 1999). Systemically, humans are not separate from, unaffected by, or free of our own limits, or the limits imposed by the systems we influence and of which we are a part. Management includes learning to function sustainably as components of not only ecosystems and the biosphere but of the interconnected universe in which we find ourselves. Precluding human existence is not an objective.

Management Tenet 2: Management must recognize that control over other species and ecosystems is impos-

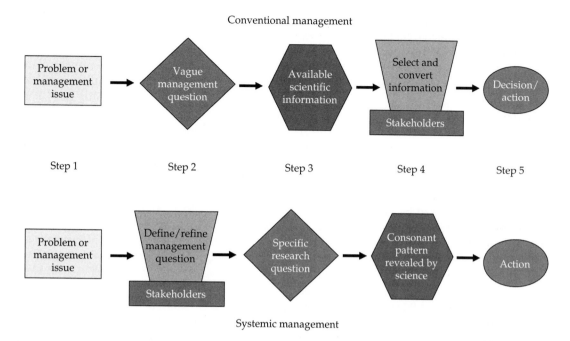

Figure 1.1 Schematic comparison of conventional and systemic management showing the difference in the role of stakeholders (Belgrano and Fowler 2008).

sible (Christensen *et al.* 1996, Holling and Meffe 1996, Mangel, *et al.* 1996, NRC 1999). The most realistic option for control is control of human action (intransitive management, Fowler 2003) and decision making—as limited as that itself may be. The self-control of systemic management acknowledges that we can control our influence on the nonhuman (e.g., fishing with all its effects on fish, and their ecosystem) much more effectively than we can control the nonhuman (e.g., other species, their ecosystems, and habitat). It is axiomatic that all human influence (like the influence of all species) has its consequences, whether direct or indirect—a fact over which we have no control.

Management Tenet 3: Management must account for the complexities of reality, including the various scales of time, space, and biological organization (Christensen *et al.* 1996, Costanza *et al.* 1992, Mangel *et al.* 1996, Moote *et al.* 1994). Systemic management is based, in part, on the reality of emergence.[2] Through emergence, natural patterns are infinitely integrative and replace human institutions in objectively and holistically accounting for complexity in the interconnectedness of nature (Belgrano and Fowler

2008). Included in this complexity is the reality of numerous human influences. Human limitations are real—there are things we will never know.

Management Tenet 4: Management must be consistent in its applications, particularly in simultaneous application at the various levels of biological organization (Christensen *et al.* 1996, Costanza *et al.* 1992, Ecosystem Principles Advisory Panel 1998, Hobbs and Fowler 2008, Moote *et al.* 1994). In systemic management, the thinking and action of individuals, governments, and institutions are consistent. Human influence on nonhuman populations or species is consistent with our influence on ecosystems or the biosphere. There is consistency between management questions and guiding information produced by science.

Management Tenet 5: Management must strive to avoid the abnormal in processes, relationships, individual organisms, species, and ecosystems (Belgrano and Fowler 2008, Christensen *et al.* 1996, Ecosystem Principles Advisory Panel 1998, Fowler and Hobbs 2002, Mangel *et al.* 1996). Abnormal conditions of any kind are the basis for taking the responsibility to find and correct underlying human abnormalities. Healthy[1] systems are an important objective of systemic management—with consistency (Management Tenet 4) involving individuals, species, ecosystems, and the biosphere. This tenet promotes the objectivity of systemic management achieved by confining the role of stakeholders to that of asking clear questions (Fig. 1.1) followed by carrying out management action.

Management Tenet 6: Management must be risk averse and strive to achieve sustainability (Christensen *et al.* 1996, Francis *et al.* 2007, Holt and Talbot 1978, Mangel *et al.* 1996, NRC 1999). Sustainability is, by definition, not achieved by any form of management that generates abnormal levels of risk. In adhering to Tenets 3 and 4, systemic management treats risk in its aggregate so that consistency is achieved. Extinction is a risk accounted for in the emergence of patterns. Sustainability is achieved by avoiding the abnormal (Management Tenet 5) as revealed in such patterns.

Management Tenet 7: Management must be based on information (Christensen *et al.* 1996, Interagency Ecosystem Management Task Force 1995, Malone 1995, Mangel *et al.* 1996, NRC 2004). Information from integrative patterns is at the core of systemic management (Belgrano and Fowler 2008, Fowler 2003, 2008). This information, at least partially cybernetic in nature, provides managers with objective, meaningful, measurable, and reasonable goals and standards (Management Tenet 9) of which maximizing information and biodiversity (Fowler 2008) is an option.

Management Tenet 8: Management must include science and scientific information (Christensen *et al.* 1996, Francis *et al.* 2007, Interagency Ecosystem Management Task Force 1995, Malone 1995, Mangel *et al.* 1996). Both conventional and systemic management rely on carefully conducted science. Systemic management, however, recognizes a difference between the products of science and the scientific process to define the best scientific information for management: there must be consonance between management questions and the scientifically revealed natural patterns used to address them (Management Tenet 4) in implementing Management Tenet 5.

Management Tenet 9: Management must have clearly defined, measurable goals and objectives (Christensen

et al. 1996, NRC 1999). Integrative natural patterns provide a basis for detecting and measuring the abnormal. The magnitude of problems before us becomes clear. Fixed-point goals are not part of systemic management insofar as patterns in variation *of variation* make adaptability part of the process. Systemic management provides consistent (Management Tenet 4) guidelines, criteria, and standards of reference for action (e.g., to maximize biodiversity; Belgrano and Fowler 2008).

Notable in the list above is the confined role of stakeholders; this is consistent (Management Tenet 4) with the requirement of objectivity inherent to Management Tenets 7 and 8. The importance of involving stakeholders is widely recognized (e.g., Arkema, *et al.* 2006, Francis *et al.* 2007, Moote *et al.* 1994, Mangel *et al.* 1996, Phillips and Randolph 2000). However, this is often equated with making human values other than that of systemic sustainability (avoiding the abnormal) the basis for the decision-making process (Franklin 1997, Grumbine 1997, Hull 2006, Mangel *et al.* 1996, Moote *et al.* 1994, NRC 2006, Norton 1987, Sagoff 1992, Salwasser *et al.* 1993) rather than simply knowing that such values count as components of the complex systems of which we are a part. Intentional use of such values in decision making effectively precludes from the tenets found in the literature a tenet specifically expressing demands for objectivity (e.g., solution to the polarity between the biocentric and anthropocentric; Stanley 1995); some claim that value-neutral approaches are a myth (Norton 2005). Systemic management acknowledges all stakeholders (managers, scientists, government leaders, organizations and institutions, indigenous societies, and individuals) and their values as parts of systems important to management (Management Tenet 1). In conventional management, stakeholders are involved in decision making (converting information to objectives and the intended course of action; Brosnan and Groom 2006) making management subjective. In systemic management, this role is replaced by informative integrative patterns (Fig. 1.1) produced through research designed in response to good management questions, posed with the help of all stakeholders. Human limitations do not enter into the process of converting information to objectives, thus dealing with the reality (Management Tenet 3) of human limits. This achieves a degree of objectivity not possible in conventional management (Belgrano and Fowler 2008). Each carefully posed management question then defines the science needed to reveal the matching pattern; the science that best meets the needs of management. Defining such science has not heretofore been possible (NRC 2004).

1.2 Motivation for change

There are a number of reasons to consider alternatives to conventional management. The need for change is seen on many fronts and falls into six categories:

• Deterioration (abnormal state) of many of the world's ecosystems, the biosphere, and the environment in general—failures of conventional management.
• Recognized inherent inadequacies that contribute to the failures of conventional management.
• Legal mandates in the form of legislation and international agreements that require managers to account for more complexity (e.g., ecosystems and other broader bioregional and global scales, evolutionary and geological time scales).
• Flaws in the traditional use of science in management.
• Lack of standards, criteria, reference points, and normative information for systems such as ecosystems.
• Lack of clearly defined, broadly accepted ecosystem-level management.

Most, or all, of the above involve human values that are brought to the task of change—change developed in this book as critically necessary. This change involves the processes of decision making, policy setting, and finding objectives. It avoids using values (other than sustainability), emotions, politics, and opinion as the basis for policy.

1.2.1 Abnormal state of ecosystems and the biosphere

One of the roles of science is to reveal problems. Science has revealed a great deal of change in the

Earth's ecosystems and biosphere (MEA 2005a, b); much of this change is anthropogenic (Moran 2006). The magnitude and pervasive nature of anthropogenic influence on our planet has reached such extremes that scientists now refer to this period of time in Earth's history as the Anthropocene Era, partly in response to the geophysical nature of some of the impacts we are having (Steffen *et al.* 2004). In many cases, scientists think of the current states of ecosystems and the biosphere as abnormal. The abnormal concentration of CO_2 currently in our atmosphere is an example (Fig. 1.2), bringing with it global warming, oceanic acidification, redistribution of water/ice, and other associated changes and ramifications yet to be recognized by science.

We have cleared forests for agricultural use and harvested timber so extensively as to cause

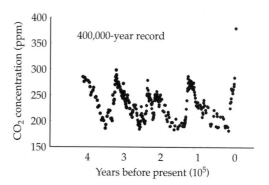

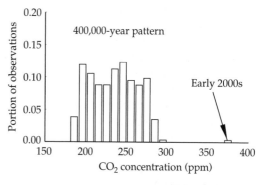

Figure 1.2 The historical record (as reconstructed from ice cores) of atmospheric CO_2 concentrations for the past 400,000 years (ppm, top panel) and the resulting pattern of variation showing the abnormality of currently observed levels (bottom panel; Dr Pieter Tans, NOAA/ESRL [www.cmdl.noaa.gov/gmd/ccgg/trends]).

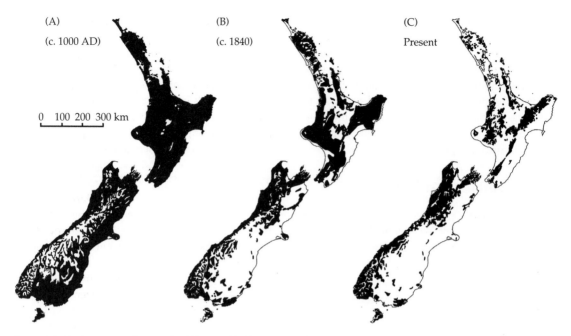

Figure 1.3 The forest cover of New Zealand (shown in black) in its decline from about 1000 AD (prior to human arrival) to 1840 (prior to European settlement) and the mid-1990s (updated from Mark and McSweeney 1990, after Froude *et al.* 1985, with maps provided by R. Philips).

dramatic declines in the area of the Earth covered by forests (e.g., Fig. 1.3). Even greater changes have occurred in grassland ecosystems (Thomas 1956). The world's commercial fisheries are largely over-harvested (even by conventional standards, Francis *et al.* 2007, Myers and Worm 2003, NRC 1999, Rosenberg *et al.* 1993) and many freshwater fisheries have disappeared (Karr 1994). Bats, bees, and other pollinating species are exhibiting population declines (NRC 2007). Many species are threatened; amphibian extinction is a small part of an overall extinction crisis (Groom *et al.* 2006, Raven and Cracraft 1999). Related ecosystem changes include: reduced food chain length (mean trophic level) and reduced mean body size in many species, increased predominance of pests and diseases, and altered species composition, distribution of biomass, productivity, decomposition, cycling of nutrients and biomass, community respiration levels, and net production. Although many problems are obvious, there is concern that we have little information about the normal states of ecosystems against which to objectively measure the abnormality (Costanza *et al.* 1992a).

The consequences of altered ecosystems to humans—especially future generations—are increasingly being recognized. One ultimate consequence is the increased risk of human extinction.[3] Short of extinction is feedback (e.g., a global pandemic, Garrett 1994) resulting in the risk of an abnormally low human population. As indicated by Rapport (1992), altered ecosystems often include the loss of goods, services in support of humans,[4] increasingly widespread disease and starvation (Pimentel *et al.* 2007), and altered coevolutionary interactions. In the larger perspective (systemically), the loss of goods and services to all species is of concern.

Although human values may not be objective, they do serve to bring our attention to factors relevant to asking good management questions. Our emotional reactions to the changes we see around us serve well to help focus on specific interconnections and relationships between humans and the nonhuman. The various values we bring to decision making can be used to focus management questions.

Among these values are those we place on healthy ecosystems. These fall within a spectrum between the utilitarian-materialistic and intangible-aesthetic. Sagoff (1992) considers these values in three categories: instrumental, aesthetic, and moral. These values have resulted in action—management that has contributed to the patterns we observe today. Current management fails to address the extent to which instrumental, aesthetic, and moral benefits can be sustainably provided by the earth's ecosystems. These benefits count among the complete set of services provided by ecosystems—important not only to humans, but to other species as well.

In other words, the goods and services recognized as important to humans are important to all species. Every species depends on its ecosystem for survival. The related services include producing oxygen, absorbing carbon dioxide, regulating climate, recycling wastes (including detoxifying pollutants), controlling pests and disease (including resistance to invasion by exotic species), pollinating plants, preventing erosion, storing nutrients, and maintaining diversity (genetic potential for the future). The loss of such services for all species counts among the risks to be accounted for in management (Management Tenet 6).

Economists find little challenge in placing monetary value on material products and many ecosystem services (e.g., see World Conservation Monitoring Centre 1992), and translate the loss of products from disturbed ecosystems to economic losses (e.g., Page 1992), whether it is a realistic translation or not. Stakeholders use these values to sway management decisions in current forms of management. As such, and owing to our part in nature (Management Tenet 1), they count among the factors (realities) reflected in the patterns we see today; these patterns are, in part, products of decision making based on the consideration of economic interests.

Less tangible and quantifiable values are behind the aesthetic and recreational uses humans make of ecosystems. Many people enjoy foraging for berries, nuts, and mushrooms, and hunting and fishing are a source of food and pleasure. Others enjoy just being in natural settings, supporting a huge ecotourism industry, backpacking/camping, nature-based healing (e.g., vision questing/fasting, and Outward Bound programs), and vacations in national parks. The value humans place on nature for nonconsumptive uses often declines to the extent it has been altered by human action. In North America, increasing value has been placed on wilderness—ecosystems in their wild state (Nash 1982). Visits by tourists to the natural settings of ecosystems seen as relatively free of human disturbance (ecotourism) have a growing economic importance for host countries and communities (World Conservation Monitoring Centre 1992).

Aesthetic value is also found in the spiritual and religious aspects of our experience of nature and seen as fundamental to what we are as a species (Nash 1982, Plotkin 2008, Roszak et al. 1995). These include the psychological effects and insight from firsthand experiences (part of the focus of the field of ecopsychology, and nature-based therapy). These effects can be lost in degraded ecosystems (Appendix 1.2) contributing to reactions in the category of "nature-deficit disorder" recognized as a lack of exposure to the nonhuman in natural systems (Louv 2005). Many would consider these spiritual/cultural losses of equal importance to the loss of more materialistic services. More objectively, such services count among the set of reciprocal relationships we have with our environment. To one extent or another, being deprived of normally functioning ecosystems and natural environments precludes our experience of normal human–nature interactions. To the extent that this happens, dysfunctional (abnormal) individual, family, or social systems may result (Gore 1992, Roszak et al. 1995). In such situations, the effect on humans of being deprived of the experience of normal forms of natural systems is analogous to the reactive attachment disorder in individual infants who are deprived of interaction with loving parents (failure to thrive or hospitalism; American Psychiatric Association 1980).[5] Part of what is involved here is our dependence on a normal state of nature to our understanding[6] of life in general (Bateson 1979), or life in all realms, from cells to the biosphere, including the psychological and psychiatric (Dell and Goolishiam 1981, Roszak *et al.* 1995).

To varying degrees, we are a species with unrealistic belief systems and socially dysfunctional patterns, in part as reflective of the abnormal in our relationships with the nonhuman. In the reciprocity of the interconnected aspects of complex systems, then, these are factors inherent to patterns observed in both society and our environment today. In this regard, the abnormal of both the human and the nonhuman is involved; they are reciprocally intertwined and part of the complexity important to management (Management Tenet 3).

1.2.2 Inadequacy of conventional management

Conventional management suffers a general lack of objectivity. In the minds of many, this might be seen as the primary problem we face in management today; in addition to objectivity as an inherent demand of the tenets of management listed above, many would list a tenet specific to this need: Management should be objective. But the inadequacies of traditional management involve much more. There is a general failure to adhere to other tenets of management. Perhaps the main problem in conventional management is its erroneous logic (Belgrano and Fowler 2008, Hobbs and Fowler 2008). We use information of one kind to make management decisions about matters of a different kind (there is a mismatch between information and policy—knowing that the pH of oceans is changing does not tell us how much CO_2 can be produced sustainably by our species). Conventional management fails to account for complexity. Current approaches are largely transitive without adequate consideration of the consequences of our attempts to control nature.

For example, problems in our continued use of conventional single-species approaches to manage natural resources are increasingly well recognized (e.g., Cairns 1986, Fowler 2003, Fowler and Smith 2004, Ludwig *et al.* 1993, Magnuson 1986, NRC 1999, Safina 1995, Schaeffer *et al.* 1988, Wagner 1977). Among these problems is the failure of conventional management to account for genetic effects

(see Conover and Munch 2002, Fenberg and Roy 2008, Law *et al.* 1993, Stokes and Law 2000, Swain *et al.* 2007, Thompson 2005), a factor recognized in coevolutionary predator-prey responses (Kerfoot and Sih 1989). Many have called for an evolutionarily enlightened form of management in view of the coevolutionary interactions (only one example of the ways things are interconnected; Swain *et al.* 2007, Thompson 2005). Management Tenet 3 requires that genetic and evolutionary factors be taken into account.

We fail to account for evolution and complexity in the conventional use of models in management. Most single-species management models assume constant parameters, while in reality such parameters change constantly, especially over long time scales (Orians 1974, 1975, Pimentel 1966). Predator-prey interactions are under-represented and even ecosystem models fail to provide holistic/realistic guidance in our use of individual resource species (e.g., Christensen *et al.* 1996, Ehrlich 1989, Peters 1991).[7] The uncertainty involved is too extensive to overcome completely and predictions are unreliable (Ludwig *et al.* 1993, Pilkey and Pilkey-Jarvis 2007). When reliable predictions are possible, most demand human evaluation and translation to management goals, which prevents objectivity in decision making.

Current forms of management all too often fail to ask clear management questions (top row, Fig. 1.1). Ill-defined management questions are posed to experts (scientific committees, advisory boards, stakeholders) who are allowed to (required to) select and use information for decision making—often making unrealistic conversions (e.g., information on production as used in finding sustainable consumption in the management of fisheries; Fowler and Smith 2004). As a result, human limitations are, by design, inherent to the process and economic factors, politics, religious beliefs, and human values serve as the basis for decisions (Hobbs and Fowler 2008). The lack of objectivity in conventional management has, as part of its origins, poorly posed management questions—specifically, questions that are referred to stakeholders rather than being used to direct science toward the research necessary to obtain objective answers.

Recognition of the flaws and shortcomings of conventional management is another step toward realizing the need to seriously consider alternative forms of management—different paradigms (Lavigne *et al.* 2006).

1.2.3 Legal mandates to extend management to include ecosystems

The seriousness of recognized environmental problems and the inadequacy of conventional approaches, have led to a variety of legal mandates worldwide.[8] Many of these require undertaking "management at the ecosystem level" or "ecosystem-based management" despite the absence of any clear, widely accepted definition for just what such management is.

In the United States, the Marine Mammal Protection Act (MMPA) emphasizes the importance of the "...health and stability of the marine ecosystem...." To comply with such a mandate, we need to know what the "health"[9] of an ecosystem is, how to measure it, and what we can do to practically apply such knowledge. As reviewed in Thorne-Miller and Catena (1991), the U.S. Marine Protection, Research, and Sanctuaries Act requires consideration of the maintenance of ecosystem structure and prohibits the degrading of ecological systems. Goals of the U.S. Fish and Wildlife Service's National Refuge System include the preservation, restoration, and enhancement of the natural ecosystems of all endangered or threatened species. The U.S. Endangered Species Act, the Magnuson Fishery Conservation, and Management Act (US Public Law 101–627) require ecosystem consideration in management. The preservation or protection of particular ecosystems is required in the management of U.S. National Parks and Refuges by a variety of statutory mandates (Thorne-Miller and Catena 1991).

Other nations have their own environmental laws and regulations, many of which require ecosystem considerations. In Europe, these involve a number of international regulations, many of which involve nations outside the European Union (Mitchell 2003).

International mandates, in general, are exemplified by the agreement to study and manage the resources of the Antarctic. The Convention for the Conservation of Antarctic Marine Living Resources explicitly requires an ecosystem perspective in obligating "...prevention of changes or minimization of the risk of changes in the marine ecosystem which are not potentially reversible over two or three decades...."[10] Other international efforts to achieve ecosystem approaches to management are found in both government and nongovernmental programs. The United Nations Environment Program places particular attention on the effects of pollution on ecosystems. The International Union for the Conservation for Nature and Natural Resources, the Biosphere Reserve Program (sponsored by the UN's Educational, Scientific, and Cultural Organization), and The Nature Conservancy all have ecosystem-oriented objectives for conservation. Many existing international agreements and treaties regarding wildlife and the environment explicitly recognize or require consideration of ecosystems in management.[11]

Complexity is important in management (Management Tenet 3). The emphasis on ecosystems in legislation and the management literature is a step toward including more complexity. But it is only a step; it recognizes the need to extend the scope of management from individuals or individual species to include ecosystems. Beyond ecosystems, we now recognize the need to include the biosphere (Fuentes 1993, Huntley *et al.* 1991, Lubchenco *et al.* 1991, Myers 1989, Vallentyne 1993). We are beginning to appreciate the importance of accounting for the complexity of all levels of biological organization from cells to the biosphere (Management Tenet 4). Accounting for ecosystems, and the legal mandates to do so, are steps in that direction. The end point of efforts to list what should be taken into account would be the full complexity of reality (Appendix 1.1)—a list that is humanly impossible to achieve in conventional thinking.

Finally, legislation in the United States requires the use of the best available scientific information in the process of management (NRC 2004). When values other than that of systemic sustainability are brought to the task of choosing information for use in management (e.g., what stakeholders consider to be "sustainable development," "sustainable growth," or "sustainable economy," Step 4, top row of Fig. 1.1), management is vulnerable to error. The erroneous nature of conventional management (Belgrano and Fowler 2008) can be viewed as just one of many major problems we currently face.

1.2.4 Flaws in our use of science

Philosophers of science have debated the utility and nature of science extensively. The limitations and imperfections of science have become a matter of principle (e.g., Christensen *et al.* 1996, Fowler 2003, Fowler and Hobbs 2002, Francis *et al.* 2007, Holt and Talbot 1978, NRC 1999). One of the main limitations of science is our inability to account for the unknown, especially in the construction of simulation models.[7] It is one thing to know we leave out crucial information (*omission of important processes*, Pilkey and Pilkey-Jarvis 2007), it is yet another to know that there are things we cannot know to include. Another limitation of science is the inability to assign objective relative importance to the various subjects of scientific study (e.g., the never-ending nature/nurture debate). We have historically taken the subjective path by relying on human judgment (e.g., the politics involved in the last step in conventional management as depicted in Fig. 1.1) to account for the unknown in science and management. We scientists find it impossible to combine our information into a coherent synthesis free of subjective judgment and almost always ignore the unknown (the Humpty-Dumpty syndrome, Fowler 2003).

Scientific inquiry often proceeds with the assumption that things that are statistically insignificant are unimportant, yet we acknowledge the importance of such factors on longer time scales in the phenomenon referred to as the "butterfly effect"(Gleick 1987), or the effects of initial conditions (Bateson 1979, Brown *et al.* 1996, Koehl 1989, Merton 1936, Pennycuick 1992, Williams 1992). Many things are too subtle to be detected by conventional scientific methodology.

Science is confined to the reductionistic; it involves human limitations, particularly the limits of the human mind. This is a reality to be accounted for in management (Management Tenet 3). Conventional decision making and management fail to find a way in which reductionism provides a path forward. There is very often a mismatch between the information used and the management question being addressed (Hobbs and Fowler 2008). A prime example is the case of using information on CO_2 concentrations in the atmosphere to promulgate regulations on CO_2 production; abundance and production are two different things. Thus, when faced with information of one kind and a management question of a different kind, conversion by stakeholders is required (step 4, Fig. 1.1) rather than avoided. This process is inherently subjective and when we choose to continue with conventional approaches, we are choosing to avoid more objective options.

Taken as a whole, science has shown us that things are interconnected (Plate 1.1) and complicated. Science helps us explain and predict; we know that there is explanation and pattern. Science has documented and explored the emergence of things that we observe. Science has discovered, named, categorized, and characterized many things. It also exposes problems. The products of science are dumped in the hands of stakeholders, managers, and teams of scientists who are asked to make sense of it all as part of making a decision (the information supplied to decision makers, step 3 in the top row of Fig. 1.1). We know that we humans are limited, yet that knowledge does not stop us from making decisions in defiance of the Humpty-Dumpty syndrome (converting and combining information selected from libraries full of available information to guide management—step 4 in the top row of Fig. 1.1). Conventional management has not found a way to use reductionistic science in a way that circumvents the problems of conventional decision making.

1.2.5 Lack of standards for ecosystems and the biosphere

Both the magnitude and number of environmental changes occurring around us are of growing concern (MEA 2005, Moran 2006, Steffen *et al.* 2004). There are exceptions (e.g., Fig. 1.2, and more qualitatively, Fig. 1.3), but in many cases there is limited basis for scientifically proving that we are seeing anything abnormal or for measuring abnormality (Management Tenet 5). We need much more information regarding normal variation for characteristics of ecosystems, especially over the long time, and large spatial scales relevant to such systems (Ricklefs 1989). Does the diversity, species composition, mean trophic level, or carbon content of (and flow through) ecosystems vary so much that ecosystems today can be considered normal? Or is such variation small enough that the changes we have caused make today's ecosystems anomalous? Identifying, measuring, and interpreting ecosystem-level change are not well-developed skills. As Page (1992) puts it: "There are no constant reference points from which functions and norms can be defined."

Regardless of how we undertake management, it is important to have benchmarks for evaluating systems (including ecosystems) for two reasons. We need to be able to identify problems whether or not we can solve them directly, and we need to know if whatever we do in management is resulting in progress. Objective assessment is a crucial element of management and there has to be a scientifically valid way of providing information on both fronts (Management Tenets 7, 8, and 9). Standards, guidelines, and reference points emerge from information on the normal range of natural variation (Management Tenet 5), but these are quite rare for ecosystems and the biosphere compared to those that are used in human and veterinary medicine for individual organisms. As the changes occur in ecosystems and the biosphere, there is more need for criteria to evaluate them.

The normal "state" of ecosystems is one thing, the normal "state" for species is another. There is growing attention to macroecology (as will be fleshed out in Chapter 2). In spite of what we have before us, there is a general lack of normative information about the influence other species have on their ecosystems and the biosphere. This impedes the comparative evaluation of human influence within such systems (Management Tenet 1), whether we currently fit in sustainably or not. Criteria are needed not only for defining and evaluating the health of ecosystems, but also for regulating our influence on ecosystems so that they achieve and

experience their own long term sustainability (i.e., are not abnormal owing to human influence).

Sustainability in the long term, however, involves change—both human and nonhuman. Developing a realistic appraisal of ecosystems depends on information for ecosystems that have characteristics and properties representative of prevailing conditions. The current state of ecosystems reflect, among other things, human influence. To achieve long-term sustainability will require systems free of abnormal human influence; we need systems with structure, function, and variability typical of normal circumstances to serve as points of reference (Norton 1987, Management Tenet 3). Natural ecosystem change dictates that we need information on variation caused by factors such as season, climate, solar dynamics, and the coming and going of ice ages—the same as needed for individuals, species, and the biosphere. A great deal of recovery time will be required for ecosystems after they are relieved of abnormal human influence. In the mean time, current information is reflective of current circumstances—including human influence, belief systems, and values.

1.2.6 Lack of definition for management involving ecosystems

Including ecosystems is part of the effort to account for greater complexity in management. A great deal of literature deals with this issue but most present principles or guidelines in indirect, vague, or general terms (e.g., Christensen *et al.* 1996, Francis *et al.* 2007, Guerry 2005, Mangel *et al.* 1996 and the references therein). Definitions are often expressed in terms of desirable qualities rather than operational prescriptions for action (Haeuber and Franklin 1996). We are faced with a variety of tenets such as the nine listed above. Historical work has resulted in a useful list of issues, principles, concepts, or processes that can be used to evaluate alternative management processes. Conventional management clearly falls short of full and consistent adherence to these standards.

The transitive and subjective aspects of the thinking behind conventional management is a major factor in preventing a good definition of ecosystem (or ecosystem-based) management. We can only manage our interactions and relationships with ecosystems. We do not fully adhere to Management Tenet 2.

However, the lack of clarity we have in defining management at the ecosystem level involves more. It involves a lack of full simultaneous, consistent, and objective adherence to all the tenets of management to include those beyond Management Tenet 2. This involves tenacious clinging to the belief that the current role of stakeholders is a viable option. In conventional management, this belief is held paramount to other tenets, especially Management Tenet 5. The need to account for or consider anything (factors, processes, principles) is translated to a human undertaking that involves thought, opinion, or nonconsonant artificial representations of reality (e.g., see Francis *et al.* 2007). Collaborative decision making, meetings of experts, and other institutional processes are used to convert scientific information to management (Fig. 1.1—a process vulnerable to politics and other human values). The fallacious inconsistency (Fowler and Smith 2004, Hobbs and Fowler 2008) inherent to this belief has contributed to the problems we face today and is a major component of our inability to define management at the ecosystem level. Anthropocentric values supersede objective consideration of the complex systems of which we are a part, of which we are composed, and of which we represent.

Our central concern is how to go about integrating the combination of information we would like to bring to bear in management. Changing the approach (Botkin 1990, Lawton 1974, Lavigne *et al.* 2006, McGowan 1990, Orians and Paine 1983, Roughgarden 1989) is less frequently seen as necessary than are improvements (better application and further development; Wilson 1998) of existing approaches. This perpetuates the problem.

1.3 Introducing pattern-based management: systemic management in the eastern Bering Sea

Systemic management relies heavily on using empirical information from other species[12] as guidance for human self-control at the species level (Hobbs and Fowler 2008). This section introduces the basic methodology of using patterns (probabil-

ity distributions, empirically derived species frequency distributions, or macroecological patterns, Fowler and Perez 1999) to replace the decision making of conventional management. Examples from the eastern Bering Sea ecosystem provide an example of full ecosystem-based management—to include management of the suite of human influences on ecosystems. The eastern Bering Sea example is developed in each subsequent chapter to illustrate not only the part of systemic management involving ecosystems but also other parts, including those involving individual species and multispecies groups or communities.

Patterns are part of the foundation for systemic management (Belgrano and Fowler 2008). Through the process of emergence,[2] they provide integrated objective information that automatically accounts for the complexity we are unable to consider in conventional practices. A probability distribution like the standard "bell shaped curve" shown in Figure 1.4 is a function (f, that varies from pattern to pattern) of all factors (e_i) that contribute to its formation (Belgrano and Fowler 2008, Fowler and Crawford 2004). Patterns integrate and account for complexity (the infinite in both panels of Fig. 1.4, or reality; Appendix 1.1) and all its elements in direct proportion to their relative importance. This includes all interactions and interrelationships—it specifically includes ecosystems. We don't need to know the e_i, or the function "f" (top panel, Fig. 1.4) or every contributing factor (arrows, bottom panel, Fig. 1.4), to use the probability distribution. It is information that represents a complete, integrated accounting of all factors through its emergence. Subjective assignment of the relative importance of various factors to be weighed in conventional decision making (Canter 1996) is replaced by the objective weights realized in nature to account for their actual relative importance. Patterns replace humans in accounting for complexity (Fig. 1.1).

Graphs of species-level patterns (species frequency distributions; Fowler and Perez 1999) are scientific representations of empirical probability distributions for various measures of species (e.g., energy consumption, population size, trophic level, geographic range size, and mean adult body size). Species are concentrated in some parts of

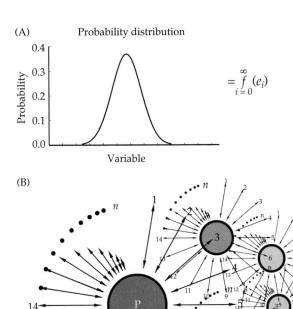

(A) Probability distribution

$$= \underset{i=0}{\overset{\infty}{f}} (e_i)$$

(B)

$n \rightarrow \infty$

Figure 1.4 Patterns as they are products of, and account for, complexity. A. Representation of typical probability distribution as a function (f) of all factors (e_i) that contribute to its formation; species frequency distributions are empirically based graphs to which a probability curve can be applied. B. Every pattern (e.g., P, in the large dark circle) is a product of, and accounts for, all factors that contribute to its formation (the infinite set of interacting factors represented by the arrows depicting reciprocity by their being double-headed).

such patterns more than others. The means, modes, and medians of the distributions indicate central tendencies that characterize the group of species represented; these are statistical characteristics of the group. The spread or dispersal of species across measures of a species-level characteristic reveals the variability among species. The limits to that natural variation are represented in the tails on either side of the curve and are measured in terms of statistical confidence limits that are also characteristic of the pattern (Fowler and Hobbs 2002). Abnormality can be identified among outlying observations.

1.3.1 Illustrating the methodology

The species with populations in ecosystems (specified geographic areas), or any other species assemblage occur in macroecological patterns (Belgrano

and Fowler 2008, Fowler and Perez 1999) as emergent structure. The full Bering Sea is a marine ecosystem located north of the Aleutian Islands, between Russia on the west and mainland Alaska on the east. For our purposes, a subsection of the ecosystem, the eastern Bering Sea, is somewhat arbitrarily defined as the area east of a line from Attu Island (western Aleutian Islands) in the south to the center of the Bering Strait in the north (Perez and McAlister 1993, see map in Fig. 1.5)—this is the ecosystem revisited in subsequent chapters. The area and its history are described in reports such as that of the NRC (1996).

The eastern Bering Sea ecosystem is of great value to humans; it provides ecosystem services to all species involved. Science has documented changes of concern. Recent commercial harvests of fish have taken millions of tons annually from a variety of resource species, and several species of

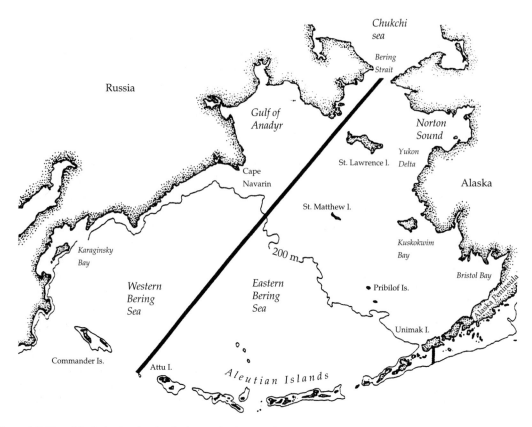

Figure 1.5 Map of the Bering Sea showing the demarcation between the eastern Bering Sea (east of the straight line) and the western Bering Sea (Perez and McAlister 1993).

marine mammals have been harvested intensively. Changes have been observed in the composition of fish species. The population of walleye pollock (*Theragra chalcogramma*) has been observed to be declining in the last several years (Ianellli *et al.* 2006). Marine mammal populations have declined and one species, the Steller's sea cow, went extinct in recorded history. These and other impacts of human activities (e.g., pollution, oceanic acidification, global warming) have generated a great deal of concern. Several species of marine mammals are subject to legal protection owing to their low numbers.

The patterns shown in Figure 1.6 for the eastern Bering Sea (based on Perez and McAlister 1993) characterize only one of the many facets of any ecosystem. This first example demonstrates variabil-

ity among the population sizes of marine mammal species based on numbers of individuals in thousands (top panel, and $\log_{10}$ raw numbers bottom panel), averaged over seasons. Figure 1.6 also demonstrates limits to natural variation in population size among species, especially as exhibited in the lower panel.

Figure 1.7 shows a different pattern with its variability and limits. It involves the same set of 20 marine mammal species as represented in Figure 1.6. In this case, we see the biomass consumed annually as estimated for the populations of these species. This graph also shows options for expressing measures of consumption. In this graph, consumption is represented from individual species (walleye pollock, top row), finfish (a group of species, second row), and the entire eastern Bering Sea

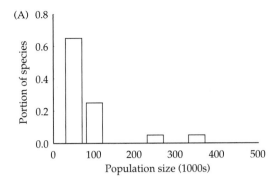

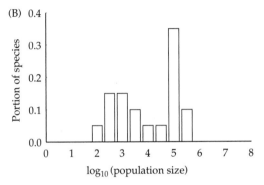

Figure 1.6 A species frequency distribution from the eastern Bering Sea, showing variability in population size among 20 marine mammal species. The raw data displayed in A are shown in $\log_{10}$ scale in panel B. (From Fowler and Perez 1999, Perez and McAlister 1993; see Appendix 1.3 for details.)

(an ecosystem, bottom row). It demonstrates two modes of presentation (raw measures and log transformed values), and gives measures of consumption rates by humans (harvests, in the right column) for comparison. Importantly, this graph introduces the kinds of data useful to the combination of single-species, multispecies, and ecosystem components of systemic management (Fowler 2002).

Fowler and Hobbs (2002) review published accounts of the basis for Management Tenet 5—the need to do what can be done to avoid abnormality. Management at the individual level is based on this tenet when action is taken to ensure that characteristics such as body temperature, blood pressure, respiration rate, food consumption, and heart rate fall within the normal range of natural variation in the practice of medicine. Applying this tenet to the harvests of walleye pollock (Fig. 1.7), a

single-species application of systemic management in the eastern Bering Sea, would mean reducing takes. The multispecies and ecosystem applications are seen in the changes indicated by the second and third rows of Figure 1.7. We begin to see the extent of both the problems and needed change: there are orders of magnitude in the differences between consumption rates by humans and the mean of consumption rates among other species. This challenge is further complicated by the need to comply with the other tenets of management. There are other management questions to be addressed.

What would sustainable harvest rates be in the absence of abnormal anthropogenic influence? Such influence (e.g., harvesting, global warming, pollution) on nonhuman species may have resulted in reduced populations. Patterns based on data for species subject to such influence would lead to indications of what is sustainable now, but may exaggerate measures of human abnormality for circumstances free of abnormal human influence. Figure 1.8, for example, shows that the biomass of cetaceans in the entire Bering Sea ecosystem in the 1980s/1990s was estimated to be about 20% of levels found there in the mid-1940s (Sobolevsky and Mathisen 1996). The same study determined that corresponding declines in population numbers (total for all cetaceans) resulted in populations in the 1990s that were about 64% of earlier levels. The most significant changes in the Bering Sea involve the extinction of the Steller's sea cow and marked reductions in sei and blue whales (the latter assumed to be represented by only small numbers of individuals in Fig. 1.8). Figure 1.8 exemplifies changes in population size and increases in the variation among species similar to changes observed in other disturbed ecosystems. These changes are part of what we take into account when we use other species as empirical examples of sustainability to adhere to Management Tenet 5. Estimates of consumption rates based on data from the 1940s, although still biased by human influence, would clearly serve as a better frame of reference for sustainable harvests from an ecosystem free of abnormal human influence—something we might achieve decades down the road. Figure 1.7 reflects human influence (e.g., values or belief systems, harvesting, pollution, global warming, politics)

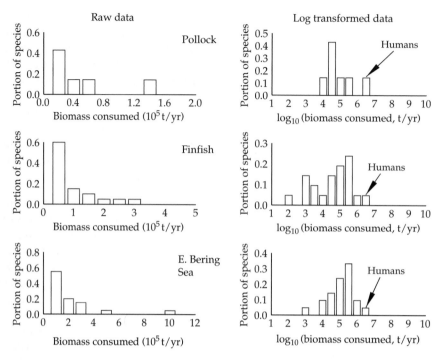

Figure 1.7 Species frequency distributions representing patterns in the eastern Bering Sea, showing variability among marine mammal species as distributed over their estimated annual biomass consumption within this region across seasons (hundred thousand metric tons, left column; $\log_{10}$ metric tons, right column) for walleye pollock (consumption by six species of marine mammals, top row), finfish (by 20 species of marine mammals, second row) and the eastern Bering Sea (by 20 species of marine mammals, bottom row). These graphs are based on information from Fowler and Perez (1999), Livingston (1993), Perez and McAlister (1993). See Appendix 1.3 for details.

along with the influence of the nonhuman (current environmental and evolutionary circumstances). This information reflects sustainability today, accounting for abnormal human influence. Free of abnormal human influence, the system may be able to regain its health to exhibit the corresponding sustainability.

1.3.2 Posing management questions

Management applied to ecosystems requires the ability to ask appropriate, but very specific, questions (Belgrano and Fowler 2008; Fig. 1.1). For the eastern Bering Sea example, we can start with questions regarding sustainability at the ecosystem level. These include:

• How much biomass can sustainably be removed by humans?
• How much biomass should be left for other species?

• How many species can sustainably be harvested as resources?
• What is the appropriate allocation of biomass removal over the various trophic levels (what portion of the harvested biomass would come from each trophic level)?

Within the ecosystem, other questions apply to harvests of individual species:

• How much biomass (or how many individual organisms) can sustainably be harvested from any particular resource species' population?
• What is an appropriate size or age composition for the harvest of any particular species (how should the harvest be allocated over size or age)?

Other questions relate to other factors both scientists and other stakeholders believe are important:

• How do we account for change and the fact that humans have caused change?

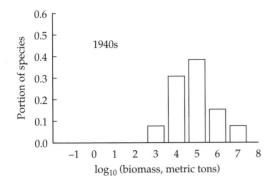

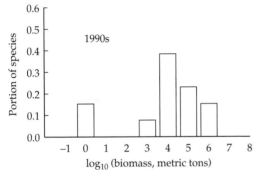

Figure 1.8 A species frequency distribution representing the entire Bering Sea, showing variability in population biomass among cetacean (whale) species (averaged across seasons, from Sobolevsky and Mathisen 1996; see Appendix 1.3 for details). This figure compares distributions for the late 1940s (top panel) and the late 1980s and early 1990s (bottom panel).

- How do we account for evolutionary impacts of our harvesting on each individual resource species?
- How do we account for the trophic level, body size, or metabolic rate of humans?
- How do we account for broad temporal and spatial scales without discounting or failing to consider the finer resolution involved in consideration of individuals, molecules, elements, behavior, habitat heterogeneity, and energy dynamics (Carpenter 2002)?
- How do we account for species-level dynamics like that involved in extinction and speciation?

Still other questions relate to human institutions and constructs in the context of the laws of nature:

- How do we account for economics in decision making?

- How do we resolve differences in management advice stemming from different sciences as conventionally applied? (For example, widely different quotas for harvests are suggested from the perspectives of population dynamics, behavior, evolution, and oceanography.)
- Does one form of science result in more reliable advice than another?
- How do we avoid ignoring or discounting any of the relevant sciences that might provide relevant information to management?

In the chapters ahead, we address questions like these. Such questions serve as effective starting points for the process of refinement; management questions are refined so that relevant issues are accounted for directly. The way we ask both management questions and science questions is very important—there must be a match or isomorphism (consonance,[13] Belgrano and Fowler 2008) between them. Initial questions must be refined to directly account for complexity. Clear questions lead to science that reveals the pattern on which management can be based. The next step is, of course, implementing what we learn from those patterns.

This is where one of the many differences between conventional management and systemic management becomes central to understanding the changes we need to make. The eastern Bering Sea again provides examples. In this ecosystem, one of the main institutions involved in management is the North Pacific Fishery Management Council. Fisheries and marine mammals are also managed under the terms established by legislation and institutions such as the International Whaling Commission. Groups of scientists provide information. Other nongovernmental organizations are involved; these include the World Wildlife Fund, the Nature Conservancy, various indigenous peoples, and groups representing the fishing industry. There are political and economic forces represented in each case. Fish are consumed by the people involved. These groups, their values and consumption are part of the eastern Bering Sea ecosystem (Management Tenet 1).

In conventional management, the debate among organizations, institutions, and individuals, such as those listed above, is a process of selecting and digesting scientific, economic, and social information to define management objectives (step 4, top row, Fig. 1.1) rather than to ask a clear management question (or questions, step 2, bottom row, Fig. 1.1). Stakeholders are doing what they genuinely believe is the best job possible in established ways of accounting for the complexity of factors needed to make a decision. Legislation allows (in many ways encourages or even requires) stakeholders to put economic values ahead of sustainability or to place primacy on economic sustainability. Such debate, evaluations, and conversion of information involves a recombination of fragmented information that is not only vulnerable to error, but ultimately impossible (involving what has elsewhere been called the Humpty-Dumpty syndrome, Fowler 2003). In the process, the bulk of reality gets ignored or misinterpreted; piecemeal tidbits of information that science provides are selectively converted to management advice. Human limitations prevent fully adhering to Management Tenet 3. Systemically, this management process is one of the factors contributing to the patterns we see; humans are part of the system (Management Tenet 1). The abnormal take of fish in the eastern Bering Sea (Fig. 1.7) is a direct product of conventional management policy—a harvest certified as sustainable by the Marine Stewardship Council (another human institution) in spite of the pathology involved in its being abnormal in comparison to that of other species. The pattern of human abnormality can be directly tied to such actions as contributing factors; denial of problems contributes to their being problems.

Systemic management puts stakeholders in an entirely different role regarding policy. Stakeholder involvement remains critical but is confined to the posing of clear management questions so that science can proceed to its task of revealing objective patterns matching the management questions (Fig. 1.1, Belgrano and Fowler 2008). Politics, economics, and human values are given objective consideration in the decision-making process based on these patterns—such factors are inherent to the patterns by being contributing factors in their emergence (Fig. 1.4). Thus, patterns account for the effects of anthropogenic factors along with the remaining set of factors important to management (the infinite of Fig. 1.4—reality). Included are evolutionary and coevolutionary forces, and other elements of the set of realities involved in the emergence of the patterns we see.

The sample patterns presented for the eastern Bering Sea in this chapter provide tantalizing insight into how systemic management works. Before delving into more detail, it is necessary to develop more about the foundation on which systemic management is based. It is important to appreciate the variety among, and within, species-level patterns and how they represent complexity. We need to understand better how patterns are natural phenomena that emerge from and reflect complexity—how they represent cybernetic information that accounts for complexity (Plate 1.2) so that stakeholders, scientists, and managers confine their roles to those in the bottom row of Figure 1.1. Then we can proceed toward a more thorough understanding of how patterns serve as guidance for systemic management. We need a deeper appreciation of the ways patterns account for human influence. With this understanding we can then explore how species-level patterns provide answers to questions such as those raised above, as well as ways to analyze patterns and choose appropriate sets of species to be most informative.

1.4 Summary and preview

1.4.1 Summary

This book draws upon work accomplished in various fields of science and integrates the work of ecologists with that of paleontologists, theoretical statisticians, complex or whole systems theorists, and philosophers of science—not to make management decisions, but to find a way in which objective guidance can be found. It introduces the practice of systemic management based on the tenets of management synthesized from the decades-long attempt to define principles for management. Conventional management has been largely transitive management *of* things like resources, wildlife, ecosystems, pests; systemic management is intransitive management of our interactions *with* all other systems (Management Tenet 2; Fowler 2003). Our management of human interactions with ecosystems is the ecosystem-based component of systemic management.

In systemic management, empirical examples of sustainability are the source of guidance, particularly as they fall into informative integrative patterns. Human concepts or constructs are confined to those most directly relevant to observing, representing, and analyzing natural patterns. The representation of guiding information is exemplified by frequency (or probability) distributions, but, for management, only those *directly matching* specific management questions. Patterns and questions must be consonant (Belgrano and Fowler 2008, Fowler and Smith 2004) to be consistent (Management Tenet 4). Patterns illustrate variation and its limits, completely accounting for all complexity (Fig. 1.4, Belgrano and Fowler 2008, Fowler 2003, Fowler and Crawford 2004, Fowler and Hobbs 2002). Frequency distributions for various species-level characteristics incorporate direct consideration of natural processes such as selective extinction and speciation (Fowler and MacMahon 1982). The examples provided by other species collectively result in a measure of what sustainability looks like (for each management question) so that we not only learn from nature, but also identify problems to be solved in management. These examples show how unsustainable the human species has become in our participation and interaction with other biological systems (e.g., other species, multispecies groups, ecosystems, and the biosphere).

In the practical applications that emerge, systemic management is demonstrated as a path toward sustainable human existence. In the end we are left with the question: can we, as a species, make the changes necessary to function sustainably?[14] Among humans, a species-level self consciousness is just now beginning to emerge. Comparing ourselves with other species has the potential of awakening in humanity both a sense of what we need to do as a species and the ability to do it.

1.4.2 Preview

The structure of the book builds toward practical application. It first sets the stage by introducing a variety of patterns and how they are formed,

especially as the selectivity of extinction and speciation contribute to their emergence. More detail concerning the flaws of conventional management are followed by in-depth consideration of systemic management as a preferable alternative, especially in terms of setting objectives. The epilogue considers systemic management with a measure of philosophy, in part addressing alternatives for action to achieve objectives. The specific example for the eastern Bering Sea, introduced above and developed further at the end of each chapter, illustrates ecosystem-based management as one part of systemic management. Single-species approaches are also part of systemic management and they too are exemplified in examples from the eastern Bering Sea as are multispecies applications. Other examples will include direct application of systemic management to the genetic effects of commercial fishing and in regard to the use of marine protected areas.

One of the underlying assumptions of this book is that human constructs show varying degrees of reliability in providing guidance for management. Among the best of what we produce are representations of empirical examples of sustainability directly matched with (consonant with) specific management questions. The more models (opinions and values) are restricted to partial indirectly related information, the more misleading they are. The most reliable guidance comes from direct observation of nature itself—as long as the observations are confined to the issue, dimensions, and logical type of the management question being asked (Belgrano and Fowler 2008).[15] This assumption is fundamental to the contrast between systemic management and conventional management; in the latter, human limitations (amplified in Chapter 4) give rise to erroneous management.

The paragraphs below summarize the focus of each of the following chapters.

Chapter 2 presents the study of variation and its limits as demonstrated in patterns among species (see also, Fowler and Hobbs 2002). Macroevolutionary patterns are introduced as emergent from the combined forces of nature, including not only human influence, but very importantly, both nonevolutionary and evolutionary processes. They demonstrate that ecosystems (or any other species assemblage) have boundaries, limits, and form—elements of structure or order with genotypic as well as phenotypic nature. These patterns are natural phenomena that represent the evolved nature of ecosystems and other natural systems to which nonevolutionary factors also contribute (Fig. 1.4); they are emergent.[2] As such, these patterns reflect the complexity that must be accounted for in management, and illustrate sustainability at the species level so as to provide guidance for human action.

Chapter 3 explores selective extinction and speciation as contributing factors in the formation (emergence) of species frequency distributions—factors that cannot be ignored in considering the complexity of reality (Management Tenet 3, Appendix 1.1). Historical consideration includes a comparison of the dynamics of selectivity among species and similar dynamics among individuals in the analogous processes of natural selection at each level, including the environmental factors that elicit selectivity. The risk of our own extinction is considered in this context. The appendices to Chapter 3 describe the mechanics of selective extinction and speciation, in combination with evolution through natural selection at other levels, to illustrate how they count among the factors contributing to the emergence of patterns in the composition of species within ecosystems and the biosphere—the characteristics of such systems.

Chapter 4 presents a detailed accounting of the problems we encounter in conventional thinking and management. Such management falls short of adhering to most of the tenets of management individually, and fails completely to adhere to all of them simultaneously, consistently, or objectively. We fail to account for complexity and we ignore the human limits that lead to this failure. Comparisons are drawn between conventional and systemic management at various levels of biological organization, and corresponding lessons, learned historically, concerning what can and cannot be done.

Chapter 5 provides more detail regarding the ways systemic management adheres to the tenets of management, objectively, simultaneously, and consistently. Pattern-based management accounts for the infinite of complexity—reality. Systemic management is reality-based management. The

implications are considered insofar as they are challenges we humans face as a species and how the potential can be evaluated in regard to what we are, and how we function, as a species. Our ways of thinking are inherent to the differences between conventional and systemic management. Thus, the chapter also looks toward the future with a brief consideration of some of the human complexity involved in changes needed for sustainability. Guidance is found by refining questions systemically to guide the science needed to provide information-based goals needing no translation or conversion. Objectivity is achieved by converting the role of stakeholder from that of information conversion in conventional management, to asking management questions in systemic management (Fig. 1.1).

Chapter 6 deals with species-level patterns as sources of information for practical application (Management Tenets 5 and 7), considering humans, along with other species, as parts of ecosystems and the biosphere (Management Tenet 1). This simultaneously allows for and requires an evaluation of ourselves and our needs as a species, by accounting for the probability of our own extinction. This chapter stresses the likelihood that conventional thinking may be one of the world's biggest problems. It is behind both a variety of risks to our own species as well as those to the world's ecosystems and the biosphere. It involves the magnitude of the human population. The related problems are numerous—for example, overconsumption, habitat destruction, extinction, and CO_2 production. If we are to avoid the abnormal or pathological through human change, it must include the systemic of human change, including our thinking on a global scale.

The book's Epilogue provides a more subjective look at the challenges of implementing systemic management and the responsibilities to be shared among humans: organizations, scientists, political and religious leaders, and all of us as individuals. While there is hope that it can occur, there is substantive basis for suggesting that it is beyond human adaptability; scientists may continue more in the role of documenting and observing the results of our management than in guiding human endeavor. On the other hand, it is possible that we examined. There are few, if any, species at very high trophic levels; food chains longer than 15 species are rare if they exist at all (Cohen *et al.* 1986).

Thus, the pyramid of numbers of individuals (fewer at higher trophic levels) is accompanied by a pyramid of numbers of species (Anderson and Kikkawa 1986), usually on a narrow base of primary producers. This is clearly illustrated by data from Schoenly *et al.* (1991), as shown in Figure 2.3. These data, from 95 insect-dominated food webs, show the frequency distribution of species within these webs according to trophic level. In parallel with other studies, the bulk (over 95%) of the species from these webs are in the lower four trophic levels.

The drop in species numbers above the trophic level occupied by primary consumers is nearly ubiquitous. Among terrestrial mammals of the southwestern United States, Patterson (1984) showed that herbivorous species outnumber insectivores and carnivores combined. Among the terrestrial mammals of North America, almost 200 (~74%) are herbivores and about 70 (~26%) are carnivores (Brown 1981). Other samples of mammalian species show the same pattern (e.g., Kelt and Van Vuren 2001) as is the case for other studies involving insects (Erwin 1982, 1983; May 1990). Such patterns are frequently observed in the literature on food webs.[4]

The explanatory aspect of science views evolutionary history as a primary component in the origin of the distribution of species across trophic level. Also, contributing to these patterns, of course, are the effects of energy depletion and availability, especially in setting upper limits to trophic level. Superimposed on this are the effects of increasing risks of extinction at higher trophic levels (Fowler and MacMahon 1982). These begin the list of factors contributing to the origin of patterns in trophic level—patterns observed and characterized by science.

How can we account for trophic level in management decisions? Management needs to ask questions such as: What is the sustainable harvest of resource species found at the fourth trophic level? How should our consumption of resources be allocated over resources at various trophic levels? How many different trophic levels should be involved in our consumption of resources?

2.1.1.3 Number of consumers

Another focus of science has been to count the number of consumer species for which a species serves as a resource. Figure 2.4 demonstrates limits to the variation in such counts. In this case, the number of predators is very limited, possibly because this sample is restricted geographically and treats only the macro-predators (i.e., pathogens and parasites are not included). Does the number of consumers per prey species change (increase or decrease) with body size? Potentially, the number of predators decreases and the number of parasites

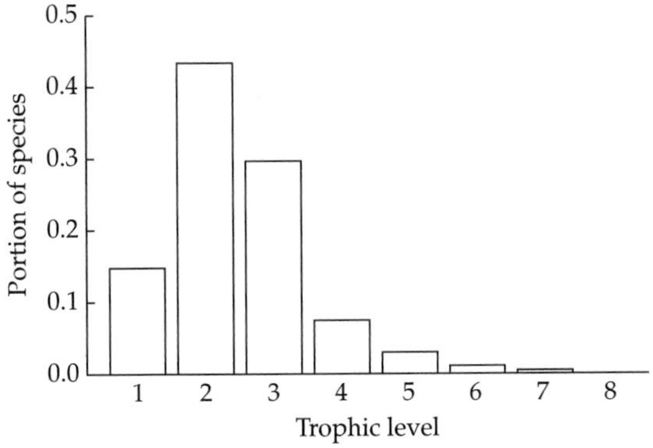

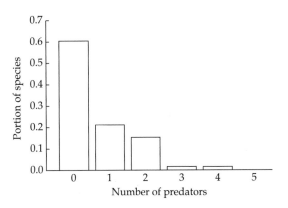

Figure 2.3 The distribution of species across trophic level for 95 insect-dominated food webs from Schoenly *et al.* (1991).

Figure 2.4 Frequency distribution of species according to the number of predators that consume them based on data for 95 insect-dominated foodwebs from Schoenly *et al.* (1991).

and pathogens increases. For example, there are at least 36 species for which the northern fur seal serves as a resource; only about three species are predators while there are 16 species of bacteria that occur internally and 17 species of parasites (Fowler 1998).

The findings of food web analysis indicate that the number of predator species per prey species is less than would be expected if predator–prey relationships were assigned randomly within a community. The number of consumer species feeding upon each resource species is a form of interaction measured as part of connectivity. The results of food web research consistently show that consumer–resource interactions are only a very small subset of the total possible interactions within any set of species. The number of predatory consumer species per resource species appears to be roughly constant regardless of the number of species in the sample or the number of species in the assemblages themselves (Cohen and Newman 1988, May 1983, McNaughton 1978, Paine 1988, Pimm 1982, Pimm and Kitching 1988, Sugihara et al. 1989, Yodzis 1980).

Science may be able to partially explain patterns such as those in numbers of predatory species per resource species, but are the explanations sufficient to address management questions? What is the sustainable harvest of a resource species already serving as a resource for eight other consumers? What is the sustainable harvest from a predatory resource species in competition with five other predators that consume a common resource species?

2.1.1.4 Interaction strength

The influence of one species on others can be measured in a variety of ways, one of which is shown in Figure 2.5 (Bascompte et al. 2005, similar to the work of Paine 1992, Appendix Fig. 2.1.4). The work of Bascompte et al. (2005), like many other cases, involves numerous cases wherein interactions involve groups of species treated collectively (as though the group is assumed to be a species). However, there are also cases of individual species interacting with other individual species. The overall pattern, despite this inconsistency, is that expected for strictly species-by-species

measurements, although its shape and the range of variation would be expected to change when interaction strength is measured for an entire species (rather than simply for individuals).

It is highly likely that the pattern of interaction strength shown in Figure 2.5 is a general pattern (similar to a normal distribution) with an intermediate maximum surrounded by fewer species toward the extremes—natural variability (Power et al. 1996) within bounds.

When we humans influence other species, at what level can we influence them sustainably? Clearly, this will depend on the kind of impact (e.g., competition, pollution, consumption, or habitat modification). Too much impact could result in reduced resources with a variety of ecosystem-level effects. When the impacts involve services upon which we depend, too little could result in an unsustainably low human population. When we ask such questions, the patterns used to address them must involve interaction measured in the same units to achieve the consistency required by Management Tenet 4, Chapter 1. Individual humans as well as our species depend on resources for their existence, and both individuals and our species have influence on other species. Here, we encounter the importance of distinguishing measures for individuals and those for species—two different logical types (individuals are parts of species).

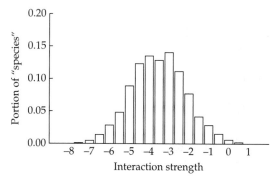

Figure 2.5 The pattern among 249 "species/trophic groups" in the strength of their interactions (log portion of prey biomass consumed per capita, $n = 3313$) with other species/groups within a Caribbean marine ecosystem (Bascompte et al. 2005).

incorporating such factors directly. Contextual circumstances involve latitude. What is the sustainable level of total CO_2 production for a mammalian species of human body size at the equator?

2.1.1.7 Allocation of consumption over space

Consumption affects both prey species and ecosystems heterogeneously over space. Consumption is allocated over space just as it is over alternative resource species, again showing variation within limits. Figure 2.13 illustrates this for the spatial allocation of resource consumption from various geographic sections of the Benguela ecosystem. Although not a trivial scientific exercise, this kind of pattern can be characterized for any ecosystem as well as the biosphere.

When our management question involves spatial distribution of harvests or consumption, a consonant empirical pattern will have information with units involving distribution. How should we allocate our harvests of biomass over different areas within an ecosystem? Figure 2.13 involves units that are consonant with this question, but only insofar as it involves the spatial distribution of consumption. The management question, however, involved human action; because humans are mammals, the consonance is partial.

2.1.1.8 Geographic range

The spatial distribution of consumption leads naturally to the topic of geographic range. Consumption of resources will be distributed over regions within the geographic range of a consuming species; areas outside the geographic range are not subject to the direct effects of consumption. Samples of data available for the geographic range size of mammals in North America (excluding marine mammals) show a pattern illustrated in Figure 2.14 for nonhuman species. In this figure, species-level geographic range size is shown as both raw measures and in log scale to show that most species are concentrated around a mode at the lower end of the range of the distribution. Brown and Nicoletto (1991) present further analysis of data for North American mammals that show the same pattern.

The geographic ranges of species overlap with each other as well as with areas that we consider to be ecosystems. The extent of the overlap can be

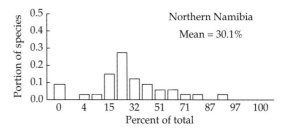

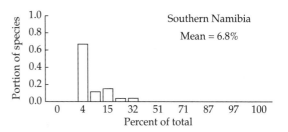

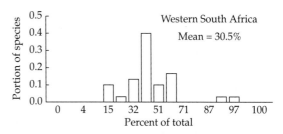

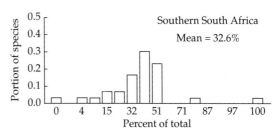

Figure 2.13 The frequency distribution of consumption (arcsin-scaled percent of total annual consumption) as allocated among four areas by 33 species of seabirds off the coast of Namibia and western and southern South Africa (from Crawford et al. 1991).

measured (varying from no overlap to complete overlap as portions from 0.0 to 1.0). Figure 2.15 shows the frequency distribution of overlaps in the geographic ranges of 21 species of marine mammals with the eastern Bering Sea ecosystem measured as portions of the eastern Bering Sea that fall within their geographic ranges. In all cases the full

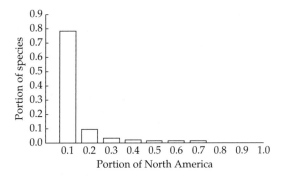

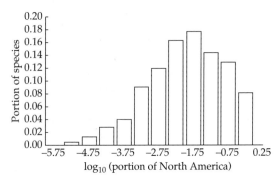

Figure 2.14 The pattern in geographic range size for 523 North American terrestrial mammal species expressed as both the portion of North American land area occupied (top panel) and in log scale (bottom panel) (modified from Pagel *et al.* 1991a with data provided by M. Pagel).

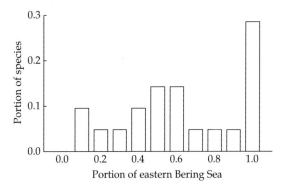

Figure 2.15 The frequency distribution for geographic range overlap with the eastern Bering Sea ecosystem by 21 species of marine mammals (from Angliss and Lodge 2004, with maps and data provided by R. Outlaw and L. Johnson).

geographic ranges of individual species include areas outside of this ecosystem (ecosystems are interconnected; Guerry 2005, Plate 2.1). There are four species whose geographic ranges include the entire eastern Bering Sea (northern fur seals, *Callorhinus ursinus*; Steller sea lions, *Eumetopias jubata*; fin whales, *Balaenoptera physalus*; and right whales, *Balaena glacialis*). Thus, the pattern shown in Figure 2.15, less one species, would apply to the overlap with the geographic ranges of all four of these species.

Total geographic ranges have been estimated for many species, including the 9505 species of birds in the study by Orme *et al.* (2006). These data, as well as data presented by Bock and Ricklefs (1983) for birds of North America show the same general pattern observed in Figure 2.14; most species are confined to small geographic ranges.

Brown (1995), Rosenzweig (1995), and Gaston and Blackburn (2000) present a broad variety of patterns involving geographic range (including graphic examples) with extensive consideration of explanatory factors and primary references. The work of Orme *et al.* (2006) and Ruggiero and Hawkins (2006) exemplifies studies of pattern—based on global geographic factors. As with all patterns, however, an exhaustive or complete detailed explanation will remain impossible owing to the complexity of factors involved. Science has made a great deal of progress, but, owing to human limitations, a full explanation cannot be achieved. From a management perspective, however, the emergence of these patterns ensures that all explanatory factors, known or unknown, are accounted for in the information they represent (Fig. 1.4).

What is the size of a sustainable geographic range? What portion of an ecosystem can sustainably be subjected to the harvest or consumption of resources? Patterns such as that in Figure 2.15 provide consonance with the latter question in representing the ecosystem in question and as information for mammalian species. The measure is of area occupied (and subject to consumption of resources)—the characteristic specified by the question. What portion of the eastern Bering Sea should be set aside in areas free of commercial fishing? To address this question, measures of area free of the direct effects of consumption would be used; the consonant pattern would be that of the portion of the ecosystem that falls outside the geographic range of each species. The two questions (how much to occupy vs. how much to leave unoccupied) would be addressed with consistent information (Hobbs and Fowler 2008).

2.1.1.9 Population density

Damuth (1987) and Schmid *et al.* (2001) present data on population density (numbers of individuals per unit area). Figure 2.16 illustrates the pattern in population density for 368 species of nonmarine mammalian herbivores (from Damuth 1987). In panel A of Figure 2.16, density is expressed in raw measurements and heavily skewed to the right (i.e., has a long right tail or is positively skewed). This is in contrast to panel B where the distribution is shown in log scale. Clearly, most species occur at the lower end of the distribution but species with

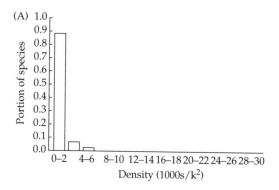

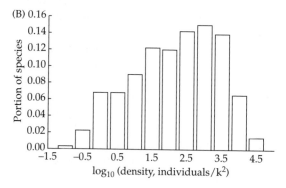

Figure 2.16 The frequency distribution of 368 mammalian herbivore species over population density, (A) expressed as numbers (in thousands) per square km, and (B) expressed as $\log_{10}$ of numbers per square km (data from Damuth 1987).

extremely low densities are rare (panel B). Related studies stress the consistency of the pattern of Figure 2.16 within taxa and various interpretations and explanations of the pattern.[6]

The distribution of a large sample of species (such as those represented in a complete ecosystem, or the biosphere) would be different from that confined to mammals. The distribution would have roughly the same shape, but would appear shifted to the right because mammals, on average, occur at lower densities than other more common, smaller-bodied species such as bacteria (Peters 1983). Gaston and Blackburn (2000) present a detailed consideration of patterns in density, especially as related to birds to show some of the variability among patterns as would be involved in the pattern for all species.

Patterns in density match management questions regarding density, at least insofar as density is concerned. If raising domestic animals is a sustainable endeavor, what is a sustainable density for sheep (measured, e.g., as sheep per square kilometer)? This question is asked in regard to domestic herbivorous mammals, and we see elements of consonance between the question and the patterns in this section—patterns involving herbivores. Other elements of consonance would be brought to the task in consideration of latitude, habitat type, soil, and seasonal precipitation.

2.1.1.10 Population size

It is well recognized that small populations are vulnerable to extinction. This fact, in combination with the information from the previous two sections suggests that we should expect to observe few species with small total populations. There are, in fact, fewer species with small populations than species with larger populations. This is seen in a compilation of estimated total population size for species of nonhuman mammals of roughly human size (Fowler and Perez 1999, Fig. 2.17).[7] Few species occur at either extreme. This sample, of course, applies only to large bodied mammalian species. Much larger total populations occur among species with smaller bodies (e.g., bacteria). Unfortunately, there are no published extensive compilations of species numbers with direct measurements of simple total population size; monitoring and assessing population size are extremely challenging

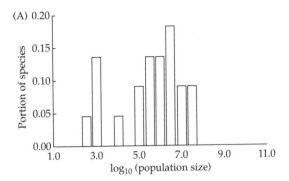

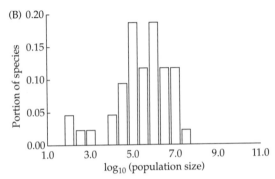

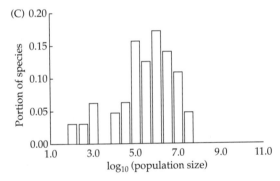

Figure 2.17 The frequency distribution of 63 species of mammals of approximately human body size as distributed over population size expressed as the logarithm of total numbers. (A) 21 species of marine mammals, (B) 42 species of terrestrial mammals, and (C) the combination of marine and terrestrial mammals in A and B (Fowler and Perez 1999, Nowak 1991, and Ridgway and Harrison 1981–1999).

scientific exercises. There is an important need for research in this regard to frame and address management and scientific questions about sustainability regarding total population size, especially for humans as will be seen in Chapter 6.

What is a sustainable population for large mammals? For any ecosystem or the biosphere, what portion of species would we normally expect to find in either tail of distributions such as that shown in Figure 2.17? The concern over endangered species leads to questions regarding the kinds of influence we have in contributing to the risks of extinction (e.g., portion of primary production monopolized, geographic range, consumption rates, CO_2 production, ...—the beginning of an essentially unending list). Implementing Management Tenet 2 (Chapter 1), "What is a sustainable population for humans?" This question will be addressed in detail in Chapter 6, exemplifying the matter of achieving specificity while maintaining consonance in the process of refining questions.

The factors that limit population size include resource availability within ecosystems in combination with the effects of predation and diseases. Much habitat is unsuitable and geographic ranges are limited. The dynamic balance that emerges from the combination of all such factors is the carrying capacity of such systems for populations. Thus, as with all patterns, patterns in population size show upper limits, and few species are expected to exhibit populations beyond such limits for any length of time—leading to the matter of variation in population size.

2.1.1.11 Population variability
Most species show little population variability compared to the maximum variability observed. This is of little surprise based on knowledge of the elevated risk of extinction associated with high population variability. However, no species is completely resistant to the effects of its environment and all exhibit some degree of population variation. Patterns emerge as shown for population variation among fish; there is a lack of species at both extremes (Fig. 2.18).

Studies of vertebrates (Fig. 2.19A) and arthropods (Fig. 2.19B) have shown a lack of species exhibiting high population variation. This is clear in Figure 2.20 for a sample of species across a variety of taxa. Although no species can withstand environmental influence to show no population variation, this is not always obvious in the ways information is presented. Choice of scale

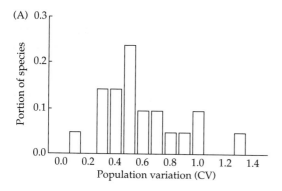

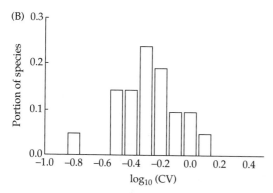

Figure 2.18 Pattern in population variation (coefficient of variation, CV) expressed in raw form in panel A and in log scale in panel B for 21 species of marine fish (from Fowler and Perez 1999, as based on Spencer and Collie 1997).

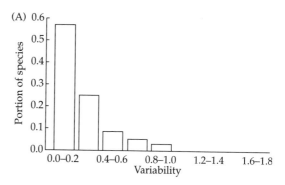

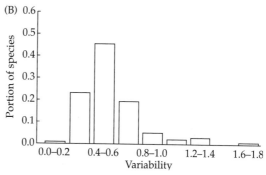

Figure 2.19 The frequency distribution of a sample of vertebrates (A) and arthropods (B) by population variability measured as the standard deviation of the logarithm of population sizes (from Hanski 1990).

is important in observing full patterns (compare Appendix Figs 2.1.8, 2.1.7, Figs 2.18, 2.19, and 2.20 to see cases where the lack of species with very low population variability is not obvious).

One meaningful measure of population variability is the coefficient of variation because of its scaled nature. However, patterns regarding population variation based on other measures are found in the literature as shown above. The magnitude of variability by some measures is related to the time over which measurements are made (e.g., longer time frames may capture more extremes, Inchausti and Halley 2001). As in all cases, observed population variation among nonhuman species reflects human influence (Anderson *et al.* 2008, Apollonio 1994, Gray 1989, Rapport 1989b, 1992, Rapport *et al.* 1985, Regier 1973, Regier and Hartman 1973, Rosenzweig 1971, Woodwell 1970, Yan and

Welbourn 1990). Observed patterns account for such influence among the many other factors involved in their origins.

Patterns in this section relate to one reaction of ecosystems to disturbance: increased population variation. How much variation is too much? How do we use abnormal variation to lead to meaningful management questions that conform to the tenets of management laid out in Chapter 1? In part, the interconnected nature of systems leads us to ask questions about human influence (Management Tenet 2, Chapter 1) and its sustainability—influence contributing to abnormal variation even if only by way of indirect ecosystem effects. We know that fishing contributes to population variation (Anderson *et al.* 2008): "At what rates should we harvest the populations of species that serve as resources?" Should the harvest of a fish species that shows high population variation be different from one that shows low population variation?

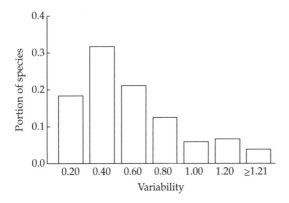

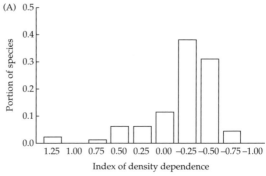

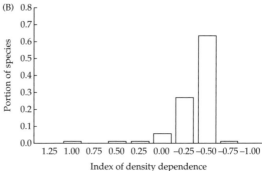

Figure 2.20 The frequency distribution of a sample of 104 species of a variety of taxa (plants, insects, invertebrates, parasites, birds, and mammals) over population variability measured as the standard deviation of the logarithm of population sizes (from Connell and Sousa 1983).

Figure 2.21 The frequency distribution of a collection of species of moths (A) and aphids (B) by level of density dependence, from Hanski (1990). Strength of density dependence increases toward larger negative numbers (reversed from usual order in these graphs to show increasing density dependence from left to right).

2.1.1.12 Density dependence

Limits to population variation are related to population regulation and density dependence—the tendency of a population to neither go extinct nor monopolize the energy and resources of an ecosystem. Figure 2.21 illustrates the distribution of several species based on the density dependence they exhibit. Density dependence in this sample involved 190 populations of 20 species of moths and aphids (Hanski 1990) measured using a method developed by Bulmer (1975). Because the variability for individual species was not presented separately from variability between sub-populations, these figures represent a combination of within-species variability as well as the among-species variability (the latter being more the focus of this chapter). Nevertheless, it is clear that most species show density dependence intermediate to the extremes.

The lack of species with high population variability is consistent with the lack of species with extreme density dependence. Extreme population fluctuations would often result from extremely positive density dependence (over-compensation, chaos, Appendix Fig. 2.1.8) with its concomitant risk of extinction. On the other extreme, lack of density dependence also carries a risk of extinction; lack of recovery from low population levels leads to species-level vulnerability. Thus, risk of extinction counts among the explanatory factors

contributing to the origin of patterns in density dependence.

What is a normal level of density dependence? For nonhuman species, this is a question for science, rather than management, as we cannot control the density dependence of other species (Management Tenet 2, Chapter 1). Among the management questions generated by abnormal density dependence would be those posed in regard to human contributions through factors which are under our control. Explicitly accounting for density dependence would be exemplified by management questions such as: What is an appropriate size composition for the harvest of individuals from a species with 0.5 as an index of density dependence (in units of Fig. 2.21)? This question defines the need for patterns involving both selectivity and density dependence to reveal what is normal and what is abnormal.

number of correlated species-level characteristics, including some that are not related to body size. Research involving correlative variation is part of the science devoted to explanation and includes characterization of patterns-within-patterns. Such information allows for the refinement of management questions and is part of the role of scientists as stakeholders in systemic management (bottom row of Fig. 1.1).

The refinement of management questions is one of the ways systemic management accounts for complexity. All patterns involve a collection of correlative interactions. Table 2.1 shows the potential for research regarding combinations of two and three characteristics (second and third columns). The fourth column shows the total of all single and multifactor combinations of potential importance in a given pattern, keeping in mind that, in reality, the number of factors involved (n of Fig. 1.4), and their synergistic interactions, are impossible to completely understand. None of the published

patterns represent an entire set of species making up a complete community or ecosystem; furthermore, studies such as that of Bascompte *et al.* (2005) always involve groups of species (higher level taxa, or trophic groups) treated as a "species" in addition to the set of species that are treated individually. The resulting graphs represent correlations better than they do the density of species within the correlative relationships (this density and its distribution being critical to defining the normal compared to the abnormal).

2.1.2.1 Body size and other species-level characteristics

Observed correlations between body size and other species-level characteristics include geographic and home range size, rate of increase, generation time, population density and size, trophic level, and metabolism (e.g., consumption and respiration). Other examples of characteristics related to body size include rates of movement, ingestion,

Table 2.1 Numbers of possible combinations of characteristics for which patterns involving species frequency distributions could be produced

Number of characteristics	Interactions between two characteristics	Interactions among three characteristics	Total number of interactions
2	1	—	1
3	3	1	4
4	6	4	11
5	10	10	26
6	15	20	57
7	21	35	120
8	28	56	247
9	36	84	502
10	45	120	1,013
11	55	165	2,036
12	66	220	4,083
13	78	286	8,178
14	91	364	16,369
15	105	455	32,752
16	120	560	65,519
17	136	680	131,054
18	153	816	262,125
19	171	969	524,268
20	190	1140	1,048,555

These are theoretical in the sense that research to actually demonstrate such interactions cannot be exhaustive, especially when numerous factors are involved as is almost always the case.

and reproduction, litter size, survival, and age of first reproduction (Peters 1983). The well-known relationship between body size and metabolism is one of the best known examples (Kleiber's rule, Kleiber 1961). Several of these are illustrated in this section (and Appendix 2.1).

2.1.2.1.1 Body size and geographic range

An example of empirical information on body size and geographic range size is shown in Figure 2.27 (from Brown and Nicoletto 1991) with the density of species represented by bars arranged in rows and columns. Each column of bars shows the portion of species in the total sample as distributed over geographic range for a selected category (span or bin) of body size. Likewise, for a given category of geographic range (row) there is a corresponding distribution of species over body size. Thus, the general patterns described for both body size (Fig. 2.1) and geographic range (Fig. 2.14) on their own are consistent with the internal patterns seen in their combination (recall the consistency required of Management Tenet 4, Chapter 1).

Figure 2.27 shows an increase in mean geographic range with increasing body size. From the other perspective, the mean body size of species decreases as geographic range size decreases. Figure 2.28 presents this pattern in different graphic form. Note the lack of cases in the lower right corner—where the combined effects of large body size and small range size result in elevated risks of extinction to prevent the accumulation of species (as at least one set of factors contributing to the observed pattern). A fitted continuous surface representing species numbers over body size and geographic range would be possible and could represent an entire ecosystem if all species were involved (see Appendix Fig. 2.1.15). The shape, but not necessarily the position, of such distributions may be expected to remain consistent from habitat to habitat (e.g., marine to terrestrial).

The combination of geographic range and body size have been the subject of numerous studies (e.g., Anderson 1977, Brown 1995, Gaston and Blackburn 2000, Gaston and Lawton 1988a,b, Hugueny 1990, May 1988, and Rosenzweig 1995).

As with other patterns in this chapter, we see the increasing potential for refining management questions. What portion of the geographic range of a species with a range size of 50,000 km^2 and body size of 50 kg should be protected by spatial limits in the harvest of one of its resource species?

2.1.2.1.2 Body size and home range size

The home range size of mammal species is correlated with body size (Fig. 2.29). This overall pattern (e.g., Appendix Fig. 2.1.13) can be broken into sub-patterns confined to particular ranges of body size (Appendix Fig. 2.1.16). Different patterns would also be seen for body size for different segments of home range size. Each such pattern is similar to those in the first part of this chapter, dealing with individual species-level traits. Similarly, each pattern demonstrates limits to variation—in this case, as related to the two correlated variables of body size and home range size. Gaston and Blackburn

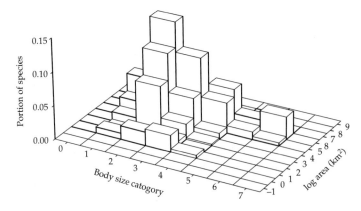

Figure 2.27 The correlative pattern between geographic range and body size for North American mammal species shown as a two-dimensional frequency distribution (from Brown and Nicoletto 1991).

rate of increase, variation in population size, home range size, and density.

Owing to the near impossibility of graphic demonstration of more than three characteristics, no figures representing the data in Table 2.2 for four or more interactions are included in this section. This small set of data alone allows for the presentation of 15 graphs in two dimensions, and 20 in three—some of which were presented above. However, there are 15 four-, six five-, and one six-dimensional patterns for this set of data for a total of 63 different ways of viewing these data if it were graphically possible to include them all.

2.1.5 Processes contributing to patterns among species

Tribute must be paid to the progress made in studying patterns among the many species-level attributes. One of the main points of this chapter is the fact that science includes the discovery of patterns. Following discovery, scientists find themselves involved in characterization, description, analysis, and explanation. Characterization

provides information on the limits in natural variation among species. Another related point is that patterns (in relationships involving one, two, three or more dimensions) define a "morphology" for the groups of species involved (as in Figure 2.34).[12] Such patterns are of particular interest in studying ecosystems, where the geographic ranges of various sets of species overlap and each ecosystem has its own set of species with patterns to be observed and subjected to research.

However, owing to the complexity of our world, the study of such patterns as far as it has progressed has barely scratched the surface. This point is made by the tiny fraction of the possible combinations (Table 2.1) reviewed in this chapter and represented in the scientific literature beyond what is covered here. Over 30 characteristics are presented in this chapter (and Appendix 2.1). With this many measurable attributes, the entire collection of species in an ecosystem could be represented by 1.074×10^9 distributions of the types reviewed above. For the collection of patterns for three characteristics (4060 combinations among the 30 characteristics) there would be correlative relationships exemplified

Table 2.2 A list of the species common to the sets of data from Damuth (1987), Sinclair (1996), and Kelt and Van Vuren (2001) showing the corresponding measures for each of the 14 species

Species name	$\log_{10}$(generation time, years)	$\log_{10}$ (intrinsic rate of increase, r_{max})	$\log_{10}$ (standard deviation, observed rate of change, r)	$\log_{10}$ (home range, km²)	$\log_{10}$ (density, N/ km²)	$\log_{10}$ (body mass, kg)
Antilocarpa americana	0.699	0.352	−0.654	3.005	−0.032	1.778
Bison bison	1.000	0.488	−0.606	4.425	−0.495	2.845
Cervus elephus	0.778	0.292	−1.051	3.874	0.739	2.279
Giraffa camelopardalis	1.114	0.260	−0.721	4.135	−0.028	2.903
Lepus americanus	0.176	0.517	0.012	0.515	2.149	0.176
Loxodonta africana	1.398	−0.019	−1.071	5.244	0.037	3.477
Microtus agrestis	−0.482	0.246	0.468	−1.155	3.342	−1.602
Mustela erminea	0.000	0.619	0.408	2.122	1.021	−0.824
Mustela nivalis	−0.301	0.573	0.492	1.812	1.556	−1.523
Odocoileus hemionus	0.699	0.047	−0.910	3.545	1.000	1.699
Ovis canadensis	0.699	0.243	−0.731	3.440	0.228	1.845
Rangifer tarandus	0.845	0.288	−0.814	5.550	0.459	2.176
Syncerus caffer	1.000	0.462	−0.724	3.912	0.580	2.653
Vulpes vulpes	0.342	0.563	−0.186	2.692	0.204	0.845

in Figure 2.34 to be seen as stereograms to better appreciate the patterns. The scientific questions to be addressed suggest the progress yet to be made, for example:

• What patterns are revealed by various three-dimensional relationships among species-level measures expanded beyond those represented in existing literature?
• How extensive is the list of dimensions over which species can be measured?
• How do patterns and relationships vary with environmental conditions (varying habitat or space: latitude, marine vs. terrestrial, etc.)?
• How do such patterns and relationships vary over time (e.g., season, climate)?

However, science does not linger on discovery or description, and explanation is never far from the minds of scientists who study relationships and patterns (e.g., Gaston and Blackburn 2000, Hubbell 2001). Patterns are natural phenomena, but what are their origins? The primary processes and dynamics that have been identified as contributing to the production of species frequency distributions are: ecological mechanics, evolutionary processes of natural selection at the individual level as well as at the species-level, primarily through selective extinction and speciation. These three sets of processes are interrelated and all operate within the context of environments as a set of conditions that contribute their own effects. The combination thus involves four domains for research[13]—unlimited science of which we already see tantalizing results. Evolutionary processes are considered in detail in Chapter 3, with emphasis on species-level processes of speciation and extinction. In Chapter 3, it is argued that speciation and extinction are more important in the formation of species frequency distributions than natural selection at the individual level, and that both are more important than the nonevolutionary factors of ecological mechanics. All operate in the context of environmental circumstances, which is increasingly recognized as subject to significant human influence. In the final analysis, all contributing factors are involved in the origin of patterns, known or unknown and regardless of the category in which they fall and regardless of the importance we think they have.

Of primary importance to the message here, however, is the conclusion that the patterns in species frequency distributions emerge[14] from the complexity of reality (Belgrano and Fowler 2008, Fowler and Crawford 2004). As such, all factors are accounted for through their contributions to observed patterns whether or not we can list them, study them, or know about them. It will be argued in Chapters 5 and 6 that while establishing the relative importance of such factors may be important in scientific debate, it is not necessary for management. This is because pattern-based management automatically accounts for the relative importance of individual factors through their contribution to the emergence of patterns (bottom row of Fig. 1.1, Fig. 1.4; Belgrano and Fowler 2008).

2.1.5.1 Ecological mechanics

Ecological mechanics (the complex of biological, chemical, and physical processes and dynamics, or the non-evolutionary processes), have been the central focus of much of conventional ecology. Ecological mechanics includes the components of which species are comprised and the interactions within and among them: populations, individuals, organs, tissues, cells, chemical compounds, and elements. Examples of these processes are predator/prey and behavioral interactions (Plate 2.2), common resource use (competition), the influence of environmental variability, pheromones, nutrient dynamics and availability, and energy flow and availability. Such dynamics and relationships involve sound (hearing), light (sight), gravity, electrical fields, and related behavioral responses. Ecological mechanics encompass many of the factors that come into play in population dynamics, all the limitations and stimuli provided by the chemical environment, and the complete temperature regime (including variation and its patterns), plus those of humidity and radiation with their variation and variety of effects.

All such factors influence the position of each species in such patterns, where each pattern represents a different expression and a different accounting of those factors (Fig. 1.4). Nonevolutionary factors help define both variation and its limits within patterns (Fowler and Hobbs 2002) and so are taken into account in management that is based on such patterns by avoiding the abnormal (Management Tenets 3, 5, Chapter 1; Fowler 2003, Belgrano and Fowler 2008).

Virtually all of the patterns presented above have resulted in examples of explanatory science following their original publication, some more than others. As seen above, it is possible to conduct an analysis of variance to decompose patterns into explanatory correlative sub-patterns. When correlative relationships involve any of the contributing factors, we can explicitly or overtly account for those factors if we have information to characterize the relationship. Such information allows for the refinement of management questions to include such factors so as to match the consonant integrative pattern (Belgrano and Fowler 2008). This includes the matter of environmental conditions. For example, Orme *et al.* (2006) show patterns in geographic range size as it varies globally among birds; questions regarding an expected range size can overtly include geographic location.

The matter of ecological mechanics is the subject of the bulk of extant ecological literature. As such, it will not be treated in as much detail as natural selection in Chapter 3.

2.1.5.2 Natural selection at the individual level

What a species is (genetically—the combination of phenotypes seen among individuals) clearly contributes to where it falls in the distribution of species within patterns. The evolutionary process of natural selection at the individual level (and among genes and combinations of genes) is instrumental in determining the genetic makeup of every species. Evolutionary dynamics involve a focus on the genetic information that overlaps ecological mechanics (natural selection often removes individuals or gene combination on the basis of related mechanical processes that don't work). Thus, phenotype or genetic design influences much of what is observed in ecological mechanics,[15] both among individuals and among species. Mechanical processes are also constrained by the raw materials making up the building blocks of life while genetic information contributes to patterns within these constraints. The constraint of context is a reality—including the context of availability of raw materials as well

as the influence of environmental variation, and a wealth of other factors.

Evolutionary dynamics have not been described to the satisfaction of everyone, partly as a result of the complexity involved. However, the predominant process is natural selection at the individual level, as we understand it through the field of evolutionary biology. Natural selection may change (or maintain) gene frequencies among individuals, and thus change (or maintain) the character and attributes of species. The resulting genetic code, the genome of every species, is a major factor contributing to what a species is and does, and where it falls in a species frequency distribution.

Evolution through natural selection operating on individuals (and the genes they carry) is linked in various ways to selective extinction and speciation, summarized below and amplified in Chapter 3—as selection operating at the species level. Part of what happens in natural selection acting on the gene frequencies within species is genetic drift (i.e., change or recombination of DNA, often characterized as random). This physical/chemical phenomenon exemplifies elements of complexity that cannot be predicted but have species-level consequences—physical–chemical processes *within* individuals *within* species. However, *among* species, genetic drift also provides an element of diffusion in species-level dynamics that contributes to the formation of variation and pattern among species.

2.1.5.3 Selection by extinction and speciation

Natural selection at the species level is another major influence on the formation of patterns among species (Aarssen *et al.* 2006, Arnold and Fristrup 1982, Fowler and MacMahon 1982, Gaston and Blackburn 2000, Slatkin 1981). A specific example of species-level selection as a pattern-shaping factor is found in the work of Damuth (2007). In this research species-level selection is examined in regard to a well-recognized pattern—a pattern in which energy use per unit area shows no correlation with body size. In simplistic terms, selection among species occurs because different species face different risks (chances or rates) of extinction and experience different rates of speciation.

In the formation of patterns, the dynamics of extinction and speciation often override those of natural selection among individuals and ecological mechanics, both of which can contribute to the formation of characteristics that make a species extinction prone (Chapter 3). Discussion and debate regarding the role of selectivity in extinction and speciation in the formation of many species-level patterns like those presented in this chapter are common in the scientific literature. Such efforts are part of the growing science of macroecology (e.g., Blackburn and Gaston 2006, Brown 1995, Gaston and Blackburn 2000).

Everything is part of something else (with the possible exception of reality as a whole, Appendix 1.1; Wilber 1995); every system is part of a larger system. The environmental context of species and the ecosystems within which they occur involves a powerful set of factors that contribute to what we see. Even the forces of extinction involve elements such as climate, island size, latitude, and continental drift. An excellent example of implementing this principle in macroecological research is that of Orme *et al.* (2006) wherein it is shown that geographic range size and geographic location show pattern. The geographic range size of species is not independent of the context provided by the earth, its continents, climate, and weather. Island biogeographic research is founded on such principles—the numbers of species in a particular area is influenced by numerous factors. These factors include the size of the area, and, as is increasingly recognized, major anthropogenic impacts.

Although other factors are undoubtedly involved in the formation of species frequency distributions, it seems clear that among the primary influences are ecological mechanics, evolutionary processes of natural selection at both the individual and species-level, and the context provided by systems in which species occur. Regardless of their relative importance, none of them can be ignored, especially in management (Management Tenet 3, Chapter 1), including those things not included in our current perceptions. In all cases, human influence counts among the factors involved.

2.1.5.4 *Context: environmental factors*

Environmental circumstances count heavily in contributing to the formation of patterns such as those illustrated in this chapter. Latitude, climate, weather, temperature, precipitation, CO_2 levels, pollution levels, and radiation are among the many explanatory factors in the origin of such patterns. Such factors not only affect patterns through the processes of natural selection, but also in their roles in ecological mechanics. The science of macroecology includes looking both at the correlative nature of patterns among species as they relate to environmental factors, but also patterns among assemblies of species. Island biogeographic research spawned major contributions to the understanding and nature of species–area relationships. The relationships between geographic range size and global location exemplify such larger-scale patterns (e.g., Orme *et al.* 2006)—patterns that can be used to refine management questions to account for what is observed by scientists.

2.2 The eastern Bering Sea example

The patterns among species displayed in this chapter illustrate measurable characteristics of any species assemblage. For example, the species frequency distribution for the numbers of species at each trophic level for the eastern Bering Sea would include every species in the ecosystem—none would be excluded. The mean trophic level, based on such data, represents an ecosystem characteristic as would the distribution of species across the various trophic levels. Thus, when expressed as portions of the species represented within the system (as described in Appendix 1.3), the distribution over trophic level (and other species-level characteristics) represents frequency, or probability, as an ecosystem characteristic. Graphic representations of the set of species with populations in each ecosystem can include all the species characteristics represented above. Such initial steps toward dealing with complexity can be extended to include combinations of characteristics, and characteristics either ignored in this chapter or yet to be discovered. Scientists recognize characteristics such as mean age at first reproduction, population-suppressing effects of predators on prey species, contributions to nutrient turnover, mean mortality rates, and mobility. The complexity of ecosystems includes the variety of ways species can be measured, including those we have not yet discovered.

Several species frequency distributions in Chapter 1 directly pertain to the eastern Bering Sea. Another, introduced in Figure 2.6, adds further information about consumption rates among the predators that feed on walleye pollock, a fish species with a population in the eastern Bering Sea. Figure 2.24 introduces patterns regarding spatial distribution of species in this ecosystem (and their direct influence on other species within their geographic ranges). The patterns illustrated for other ecosystems (e.g., Figs 2.7, 2.13, and Appendix Fig. 2.1.6) are the kinds of patterns that also exist in the eastern Bering Sea. Such patterns can be described for any ecosystem, notwithstanding the considerable logistic difficulty in the research required to do so.

As mentioned briefly in connection with some of the patterns above we can take further steps in refining management questions. What is a sustainable harvest of walleye pollock for our species as a large mammal functioning as a predator in the eastern Bering Sea? This is a management question. Although relevant, especially as they involve factors that are relevant, most of the patterns earlier in this chapter fail to match this question. Consonance between management question and integrative informative pattern must involve similar circumstances. Out of all of the figures reviewed above, only Figure 2.6 is for consumption of pollock in the eastern Bering Sea. It fails to be fully consonant with the management question, however, in that it includes estimates of consumption rates for birds, whereas we are mammals.

Consonance between question and pattern must involve identical units. It must involve the same categories as in the case of dimensionless numbers like the rate of increase per generation time (Fig. 2.25). If commercial harvests are measured in tons of biomass per year, Figure 2.9 does not provide what we are after because it involves the measure of numbers or portions of standing stock biomass. The information in Figure 2.9 does involve consumption, however, and our management of harvests measured in mass must be consistent with this information (Management Tenet 4, Chapter 1). How do we account for the body size of prey? Figure 2.6 involves walleye pollock themselves, prey size is inherent to the pattern.

What is a sustainable harvest of walleye pollock that accounts for our species as a large mammal functioning as a predator in the eastern Bering Sea? What should that harvest be to account for global warming, pollution, a history of commercial whaling, and the effects of commercial fishing on other species in this system? Figure 2.6 represents data for species with positions within the pattern as products of the effects of such factors. Such effects are inherent to the pattern. There may be continued reactions to such factors, and patterns reflective of more recent conditions may better account for what is currently sustainable. As the systems change and react, sometimes with considerable time lags, to environmental circumstances, different patterns may emerge. Although there may not be an explicit representation of the effects of past human activities, these factors are accounted for by the observed pattern; the species involved have shown response to such factors.

Ecosystems such as the eastern Bering Sea can be characterized in all their complexity by the species-level patterns within them, dimension by dimension, much as we characterize individual humans by measures such as cholesterol levels, body mass, heart rate, intraocular pressure, blood pressure, and body temperature. Patterns among species demonstrate limits to natural variation. Collections of means, or other measures of central tendencies, among these patterns are the basis for characterizing variation among ecosystems or within any ecosystem over time. This, in turn, helps characterize patterns at the ecosystem level enabling ecosystem assessment as suggested by Rapport (1989a, b) and others.

In implementing Management Tenet 2, Chapter 1, however, we can't fool ourselves into thinking that we can control ecosystems. Ecosystems (including the fact that ecosystems interact among themselves, Guerry 2005), involve too much complexity to avoid the unintended consequences of attempts to control them. When we observe lack of health in an ecosystem (e.g., an ecosystem that shows a trait outside the normal range of natural variation among ecosystems) we can ask management questions regarding the various ways we (humans) might be contributing. If we are involved in abnormal influence in our relationships with the biosphere, any ecosystems, or their components, we can then relieve those systems of the anthropogenic problem(s) as our contribution to solving any observed problems.

2.3 Summary and preview

Management is human action so as to retain its intransitive quality, as introduced in Chapter 1. In this chapter, we have seen that there are patterns among species representing ecosystem structure and function reflective of the complexity we want to account for in establishing goals for management. We can measure the abnormal among species, based on these patterns, and we can measure the abnormal among ecosystems in comparing patterns. When humans are the abnormal among species, there is basis for management action, especially when it involves our interactions and relationships with other species, ecosystems, or the biosphere. With this foundation for management, it is clear that the best available scientific information for setting goals in management are the patterns that match management questions so that action can be taken to avoid the abnormal (Management Tenet 5, Chapter 1). The science that best serves management, then, is research that brings information about those patterns to managers and other stakeholders. These are people—people involved in asking the management questions (bottom row of Fig. 1.1). The questions they ask define the kind of information patterns must contain. In this chapter, the asking of such questions was introduced (and will be exemplified in more detail in Chapter 6).

Science does more than discover and characterize patterns, however; it also involves explanation. This aspect of science develops an understanding of some of the elements of emergence behind patterns. More holistically, this aspect of science leads to the understanding that when patterns consonant with management questions are used to guide management, the elements of emergence (of reality, Fig. 1.4) are taken into account. Of the processes contributing to patterns, nonevolutionary factors and natural selection acting on individuals are relatively well understood. We know that the environment provides contextual constraint and influence on such factors—acting to prevent the abnormal. However, only in the last several decades have scientists paid much attention to the influence of selective extinction and speciation in the formation of macroecological patterns—especially in preventing the accumulation of examples of species beyond the extremes revealed by patterns. In conventional management, extinction and speciation are largely ignored, especially insofar as they are selective; when they are considered it is through the error prone processes of conventional management (top row, Fig. 1.1). In this context, Chapter 3 considers extinction and speciation in more detail, particularly regarding the limits made obvious in patterns—limits with which we can recognize the abnormal or pathological. Risks and limits are part of what we are required to consider in systemic management (Management Tenet 6, Chapter 1). Understanding them emphasizes that extinction is a risk faced by our own species.

or mentioned repeatedly (e.g., Grant 1989, Rensch 1959, Stanley 1979). His approach to natural selection is much broader than represented in Neo-Darwinism. The following are quotes from "The Origin of Species" (Darwin 1896):

...Natural Selection almost inevitably causes much Extinction of the less improved forms of life....

This great fact of the parallel succession of the forms of life throughout the world, is explicable on the theory of natural selection. New species are formed by having some advantage over older forms; and the forms, which are already dominant, or have some advantage over the other forms in their own country, give birth to the greatest number of new varieties or incipient species.

The extinction of species, and of whole groups of species, which has played so conspicuous a part in the history of the organic world, almost inevitably follows from the principle of natural selection....

Natural selection, as has just been remarked, leads to divergence of character and to much extinction of the less improved and intermediate forms of life.

Hence, the more common forms, in the race for life, will tend to beat and supplant the less common forms, for these will be more slowly modified and improved. It is the same principle which, as I believe, accounts for the common species in each country...presenting on an average a greater number of well-marked varieties than do the rarer species.

Extinction has only defined the groups: it has by no means made them....

...species which are the most numerous in individuals have the best chance of producing favorable variations within any given period.

Chapter IV of Darwin's *Origin* contains a general treatment of the process, and an entire section entitled "Extinction" includes a strong flavor of differential extinction probabilities dependent on the characteristics of species. The opposing forces of natural selection at the level of the individual and that for species are recognized. Darwin clearly understood the concept of patterns among species as being a product of selective extinction and speciation. He also considered speciation as the production of the raw material upon which extinction operates.

As reviewed by Brooks (1972), the work of Lyell in his *Principles of Geology* (published between 1830 and 1833) demonstrated that extinction was important in the dynamics of the numbers of species. Wallace's and Darwin's work added the concept of

the origin of species to supply the raw materials upon which extinction could act and to prevent extinction from obliterating life. The early evolutionists, however, emphasized taxonomic groupings of species, and were less concerned with patterns of significance for sets of species such as those from which ecosystems are assembled. Wallace and Darwin did little to develop explanations for the form and function of groups of interacting species or ecosystems other than for the nature of the species themselves. The main focus for both men was the manner in which species changed and arose, especially regarding taxonomic affinity, an emphasis that still dominates much of biological thought.

3.5.2 From Darwin and Wallace to 1980

In modern ecosystem-level biology, less progress has been made in incorporating the concept of selective extinction and speciation than in evolutionary biology and paleontology (where there also remains considerable room for progress: Jablonski 2007). The inertia of this situation has its roots in accomplishments prior to 1980. Recent reviews of material related to selective extinction and speciation point to a great deal of scientific literature prior to 1980. A brief introduction to the progress and related literature is helpful.

Useful accounts of the developments of this period are found in Arnold and Fristrup (1982), Brooks (1972), Eldredge (1985), Fowler and MacMahon (1982), Grant (1989),[39] Okasha (2006), Salthe (1985), Slatkin (1981), Stanley (1979),[40] and Williams (1992). A critique of macroevolution and its history by Hoffman (1989c) is also a valuable source of information. Of course, the literature prior to 1980 also contains useful historical information (e.g., Bateson 1972, 1979, Gould and Eldredge 1977, Hull 1976, Lewontin 1970, Stanley 1979, Wynne-Edwards 1962). Early in the last century, mention of selection at the species level is found in De Vries (1905). The evolution of ecosystems through natural selection operating among a variety of levels of biological organization, including species, is a concept appearing at least as early as 1949 (Allee *et al.* 1949).

The first appearance of selective extinction and speciation in general textbooks may have been the brief mention of interspecific selection in Simpson and Beck (1965). The next general text to include the concept may have been that of Futuyma (1979) who states: "…group selection in the form of species selection must certainly operate; a community contains only predator species that are not so efficient that they eliminate their prey and thereby themselves." (Note that species selection is equated with group selection.)

The late 1970s may be characterized as a period during which acceptance and recognition of selective extinction and speciation became established, if no more than as a debatable concept (e.g., Eldredge and Gould 1972, Gould and Eldredge 1977, Stanley 1975a, 1979, Vrba 1980). Lewontin (1970) and Hull (1976) provide early descriptions of the general process of natural selection and mention its applicability to species as the units subject to differential survival and reproduction.[41]

The early development of the field emphasized the selectivity of extinction more than speciation (Arnold and Fristrup 1982). Lewontin (1970) and others presented elements of the idea in conceptual form and certainly allowed for the speciation component. However, people working with empirical data, and paleontologists in particular, tended to dwell on extinction (e.g., Eldredge and Gould 1972). By the mid 1970s this began to change (e.g., Arnold and Fristrup 1982, Gould and Eldredge 1977, Stanley 1979). Stanley (1975a) clearly included speciation, stating that the process "…favors species that speciate at high rates….". Today the speciation component is clearly defined, and understood to be a primary component as described above.[42] It should be clear that this is a matter of conceptual history and may have little bearing on the relative effects of selective extinction in comparison to selective speciation in determining the characteristics of species-level patterns in reality.

The palaeontological and, to a large degree, the evolutionary literature dwell on genealogy or taxonomy in the treatment of selective extinction and speciation rather than on its implications for ecosystems or biosphere. Macroevolution, primarily a product of palaeontological work, is represented in the literature essentially as selective extinction and speciation applied to phyletic groups. As such, the sets of species to which selective extinction and speciation apply in macroevolution are taxonomically related groups of species, or clades (Cracraft 1982, 1985b, Damuth 1985, Greenwood 1979, Levinton 1983, Vrba 1984). Macroevolution has been defined as "…evolution above the species level, evolution of the higher taxa, or evolution as studied by paleontologists and comparative anatomists…." (Mayr 1982). Stanley (1975a) says: "…*species selection* operates upon species within higher taxa….". As Gilinsky (1986) expresses it: "Species selection…is a theory for explaining patterns of differential representation of species of certain kinds within clades; that is, it is a theory for explaining *taxic* trends".

Ecosystem-level implications of selective extinction and speciation received little attention prior to 1980. Ecosystems have been rarely considered, in the history of ecosystem science, as units affected by the evolution of the sets of species from which they are assembled, especially prior to 1980. Golley (1993), Hagen (1992), and McIntosh (1985), for example, presented histories of the ecosystem sciences in which essentially no mention is made of selective extinction and speciation. The concept has been given little or no credit for contributions to understanding or providing a basis for practical management in regard to ecosystems—a situation that continues. The bulk of ecosystem science has been, and continues to be, descriptive in this regard. The identification of species level characteristics of importance to selective extinction and speciation has been the critical, but largely unwitting, contribution of this work.

The lack of integration of selective extinction and speciation into ecosystem science did not preclude the development of its specific elements in more theoretical ecological settings. Island biogeographic theory is a prime example. Early work in this field (e.g., MacArthur 1972, MacArthur and Wilson 1967) is directly relevant to selective extinction and speciation in the examination of extinction as related to geographic range size. As a result of these efforts, species-specific geographic range size is one of the most studied and well-understood dimensions in its effects on risk of extinction.

In the 1980s and 1990s much useful research in the field of island biogeography contributed to the more general concept of selective extinction and speciation. Island biogeographic work was expanded to include consideration of a number of species characteristics that influence the probabilities of extinction. For example J. Brown (1971, 1981) suggests that in addition to geographic range size, factors such as trophic level, body size, and specialization play roles in differential extinction. The treatments of Brown (1995) and Rosenzweig (1995) exemplify the expansion and growth of island biogeographic work.

3.5.3 1980 to the present

The recent history of the development of selective extinction and speciation is far too voluminous to adequately present in the confines of a chapter (see the account in Okasha 2006). Since 1980, two major groundswells of work seem to be merging. This is the merger of the evolutionary and the conventional approaches to the study of ecosystems; studies are no longer confined to a focus on ecological mechanics (although the same cannot be said about the role of the related science in policy and management). The merger of evolutionary and ecological dynamics in science has advanced substantially through addition of natural selection at the species level to the evolutionary component. Many conventional life sciences serve to identify characteristics of species over which natural selection among species operates. The results are information-based approaches exemplified by Eldredge (1985), Salthe (1985), and Williams (1992). Conventional ecosystem sciences are described in historical reviews by Hagen (1992) and Golley (1993).

Evidence of the merger of conventional and evolutionary views is found in the work on both approaches. In conventional ecosystem sciences, for example, the processes of natural selection at the species level are discussed by Allen and Hoekstra (1992) and Allen and Starr (1982). In paleontology, the work of Eldredge (1985) is a good example wherein the merger exhibits significant development. The bridging of these approaches, for example by considering selective extinction and speciation as contributing factors in the evolution of biological communities and ecosystems, is a critical step leading to the formulation of systemic management (Fowler 2003, Fowler and MacMahon 1982). Philosophical consideration of this merger is also found in related literature (e.g., Salthe 1985).

From Wallace and Darwin to the present, there has been an increasing tendency to focus on specific elements of selective extinction and speciation. Such specificity is often reflected in the terminology (Appendix 3.6). Much of this work is done, and the terms are used, without mention of the broader context. This seems particularly obvious in the separation of the fields of paleontology and ecology, a separation that is less prominent in work exemplifying the union between ecology and evolutionary studies involving natural selection at the species level realized in recent decades as mentioned above.

Consideration of the species attributes that result in variability in extinction and speciation rates are found in both ecological and palaeontological literature. Examples with palaeontological focus are numerous (see references in Appendix 3.1).[43] These are directly tied to the extinction and speciation processes. Ecological and ecosystem research, meanwhile, has resulted in the accumulation of species-level patterns with an increasing tendency to mention the link to selective extinction and speciation. The work of Gaston and Blackburn (2000) exemplifies this trend. One of the examples of direct and overt implementation of the concepts of selective extinction is that of Damuth (2007) who considered the lack of correlation between body size and energy consumed per unit area.

Part of the progress seen in the merger of studies of ecological mechanics and evolutionary paradigms in recent years is the recognition that the character of various sets of species develops over time (i.e., have their own form of ontogeny). Damuth (1985) and Grantham (1995) point out that, historically, the unit that evolves through selective extinction and speciation has received relatively little attention. Fowler and MacMahon (1982) identify such units among the sets of species from which communities and ecosystem are assembled. This step, however, is only part of the larger context wherein progress in recent years has linked the variety of hierarchical levels of units that evolve and units of selection (Williams 1992). Progress at the ecosystem level is an example of the linkages between the material- and information-based views of biological systems.[44]

Modeling has been central to conventional studies of the ecological mechanics of ecosystems (Golley 1993, Hagen 1992). These efforts have largely been attempts to mimic ecosystem characteristics, but usually fall prey to the fallacy of composition where it has been assumed they are adequately explanatory.[45] By design, they involve *omission of important processes* (Pilkey and Pilkey-Jarvis 2007) by omitting evolutionary dynamics and interrelationships. Alternative models have demonstrated explanatory potential and see ecosystems as assemblies of populations from sets of species with frequency distributions influenced by selective extinction and speciation. Such efforts have occurred in recent years amidst steps to include more of the various elements of selectivity. Slatkin's (1981) work is one example. Extinction and speciation were included in his model "by assuming some dependence of either speciation rates or extinction rates on the phenotypic character". This model also included directional speciation as a bias in phenotypic change.

The larger concept of selective extinction and speciation is emerging in the recognition of the simultaneous realities of both ecological mechanics and evolutionary processes operating to influence various sets of species. In essence, for ecosystem modeling, this is an amalgamation of historical dynamic systems approaches and the information-based approaches embodied in models like Slatkin's (1981).

Progress in the merger of mechanical and evolutionary process is evident in the work of both paleontologists and ecologists (e.g., Allen and Hoekstra 1992, Allen and Starr 1982, Brown 1995, Damuth 2007, Eldredge 1985, Gaston and Blackburn 2000, Jablonski 2007, Kitchell 1985, Rosenzweig 1995)[46]. But general textbooks seem to lag behind in this regard. In the second edition of Futuyma's book on evolutionary biology (Futuyma 1986a), the term "species selection" is defined to occur when "speciation rates or extinction rates of whole species are effected by whether or not they possess a particular characteristic; consequently, the proportion of species in a higher taxon that bear one trait or another might change". Note here that the process

is conceptually related to taxa as the pertinent set of species. Begon *et al.* (1990), in a general ecology text, briefly mention the concept of selective extinction and speciation. In an otherwise predominantly descriptive approach to the topic, the authors of this text draw a connection among ecosystem properties, species frequency distributions (composition), and selective extinction and speciation. Enger and Smith (2000) provide a slightly more detailed account of selectivity and present a short list of characteristics that lend to extinction risk. This text, like most, however, does not link selectivity to the evolution of the sets of species making up ecosystems.

Maynard Smith (1983) reduces selective extinction and speciation to a single sentence in saying "…species will differ in their likelihood of extinction, or of further splitting, and these likelihoods will depend on their characteristics". This form of natural selection is occasionally seen as part of natural selection as it operates at all levels (Jablonski 2007, Okasha 2006). The resulting frequency distributions of species are seen by Fowler and MacMahon (1982) as ecosystem properties in the form of structure and function acquired through selective extinction and speciation operating on the species from which ecosystems are assembled. The characteristics acquired through these processes are influenced by the nature of the abiotic environment. This is a role of selective extinction and speciation that contributes to the explanation of why ecosystems differ from one physical environment to another. The concept that sets of species adapt to their abiotic environments as an aspect of selective extinction and speciation is yet to be developed in a useful (applicable) way.[47]

3.6 Implications for management

What does all of this mean for management? From the practical perspective, patterns among species embody the effects of evolutionary factors, including selectivity at the species level. As sets of information, the patterns themselves account for such processes along with all other contributing factors. This is the importance of Figure 1.4. In the voluminous literature on management, strong emphasis is placed on the importance of accounting

• We manipulate systems to proceed as though we have control where we do not.

• We are far from achieving consideration of complexity, especially in inadequate considerations of ecosystems and the biosphere.

• There is little if any consistency in application at the various levels of biological organization.

• Abnormal situations abound.

• Risks are among the factors of complexity not adequately woven into management.

• Management decisions are not fully informed by anything like a complete consideration of complexity.

• Science is not used for its strengths, and we fall prey to its weaknesses.

• Many goals and objectives involve subjective guesswork.

All such problems are failures to adequately adhere to management principles developed in the Airlie House[31] meetings (Holt and Talbot 1978, Mangel *et al.* 1996) and those developed in considerations of "ecosystem management" in recent scientific literature (Appendix 4.1, Fowler *et al.* 1999, Fowler and Hobbs 2002, McCormick 1999). Efforts such as the Airlie House meetings, the work sponsored by the American Ecological Society (Christensen *et al.* 1996), and many other efforts (Arkema, *et al.* 2006, Ecosystem Principles Advisory Panel 1998, Fowler 2003, Francis *et al.* 2007, Lackey 1998, McCormick 1999) have had the expressed purpose of bringing together scientists and experts in "natural resource management" to identify the fundamental aspects of the management process. The common patterns are embodied in the nine tenets of management introduced in Chapter 1 (noting again the exception of the nature and timing of stakeholder involvement), fleshed out in the literature mentioned above, and left standing after exposure of the flaws of the transitive in conventional management. These failures were described at the beginning of this chapter and illustrated in more detail in consideration of focal level management above.

Primary among goals consistently identified for management in regard to resource use has been the maintenance of resource systems in states that can provide sustained benefits and exhibit resilience in the face of environmental and anthropogenic change (Appendix 4.2). This involves the balance between human needs and the capacity of systems to meet them such that both human, other species, and resource systems exhibit normal properties.

4.4.2 Attempted control ignores repercussions

Management Tenet 2 emphasizes that we may influence but not control other systems. Interconnectedness is much more prevalent than we can deal with in conventional management; we cannot avoid secondary or indirect effects of our attempts to control (Fowler 2002, Fowler and Hobbs 2002). Attempts to control individuals has taught us that certain changes, behaviors, or levels of production are not possible; we cannot make an Olympic athlete out of everyone. Undesired changes often result from attempts to control, manifested as side effects, secondary effects, or cumulative effects. Employees resign, pets die, and stresses express themselves in various pathologies. The lack of control is even more obvious in managing species, as exemplified by evolutionary reactions, such as changes in gene frequency, and life history characteristics. Influence brings about changes within the systems in which managed species are represented, changes that are unpredictable and often of negative consequence to humans, including the managers. At the level of ecosystems and the biosphere, belief that control is possible is probably at the heart of impasses in carrying out successful management. Changes in reaction to our influence are beyond our control, especially evolutionary/coevolutionary changes, which are further complicated by ecological mechanics. Because of the complexity involved in interactions such as consumer/resource relationships and consumption of common resources, it is impossible to predict outcomes in other parts of the system and even more impossible for changes at other levels of organization. The resulting changes include feedback as consequences of direct importance to us humans.

Underlying the concept of transitive management is the fundamental assumption that we have control over our environment, an assumption that is refuted by the problems we face. Attempts to control ecosystems result in changes. Some may be to the advantage of individuals and cause problems for species. Some may benefit species but present risk to individuals. The complex, unpredictable, changing nature of living systems dictates that we do not have the control over other elements of life necessary to practice transitive management over long time frames. Our ability to control is probably impossible for ecosystems (Mangel *et al.* 1996) and the biosphere, highly unlikely at the species level, and difficult at the individual level, even when the individual and the manager are the same human individual.[32] In the final analysis, we have maximum control over ourselves.[2]

Now we must face the distinction between what is possible and what is not. Identifying when, where, and how control (Plate 4.2) is possible is important at all focal levels of management. Fundamental to this is recognizing that managers almost always have more control over their impacts and influence on a transitively managed system than over the nature of, or process within, the managed system. Systemic management uses self-control, which emerges as the common element of success in management regardless of the level of biological organization.[33] Lack of control is clearly exemplified in the genetic effects of human influence on other species.[34] This leads to the question of how to exhibit control at the human level so as to achieve objectives at all levels of biological organization simultaneously.

4.4.3 Conflicts are irresolvable

So far, resolution of the conflict inherent in the interplay of managing different levels of biological organization simultaneously has not emerged in conventional transitive approaches, in violation of Management Tenet 4. Conventional approaches fail to prescribe an appropriate balance among individuals, species, ecosystems, and the biosphere in management.[35] We cannot expect otherwise because of: (1) the opposing forces between different hierarchic levels,[36] (2) our ignorance of complexity, and (3) the related dynamics of emergence. Long- and short-term, broad scale and local, and ecosystem and individual objectives are often in conflict. Recognition of such conflict helps identify another major stumbling block in conventional management: there is no means of finding balance

or compromise. Intransitive management can fill this gap. It adheres to all 9 tenets of management as fleshed out in Chapter 6; it solves most of the problems identified above.

4.4.4 Vulnerability to the fallible, finite nature of humans

It hardly bears repeating, but the importance of the matter is at the core of the problems of conventional management: we humans are finite, fallible, and biased, whether as scientists, politicians, religious leaders, or environmentalists. In principle, this is well understood but not fully taken into account in our choice of decision-making processes; we continue to choose the top row over the bottom row of Figure 1.1. Humans are thought to be of central importance in decision-making as defined by the role of stakeholders in conventional management. In this role, we choose, interpret, and convert partial information to management objectives. As embodied in conventional management this process epitomizes the problems described by Dante in his epic "Comedy" (later renamed Divine Comedy, Meeker 1997)—using fragments of the whole counterproductively. It involves misusing piecemeal knowledge combined with human values—eating of fruit from "the tree of knowledge of good and evil" as suggested to be fatal in the Bible (Bateson 1972). It is not holistic in the root sense of the word holy; it does not involve "thinking like a mountain" (Leopold 1949).

The arrogance and hubris of this problem has been the subject of many scholarly works, yet we continue with this as our way of deriving goals, policy, and objectives for management. We even see it to be so important as to train specialists to undertake the conversion process (Brosnan and Groom 2006) rather than understanding that it is an example of intellectual alchemy (converting kinds of information; Belgrano and Fowler 2008). Finding consonance between management question and empirical pattern (bottom row of Fig. 1.1) is not an option in conventional management. As will be seen, finding consonance is not a matter of rejecting the reductionistic quality of science and human thought as much as it is realistically picking the reductionistic information so as to avoid misuse—getting a match between management question and scientific information (Belgrano and Fowler 2008, Hobbs and Fowler 2008).

4.4.5 Belief systems deny the gravity of these problems

People involved in conventional management typically share beliefs (or assumptions) that:

• Using current approaches to resolve conflicts is the best we can do; inconsistency is a problem to be endured, not solved.
• Existing ways of considering complexity suffice; there is really no compelling need to consider alternatives, especially if they require difficult change and sacrifice.
• Existing ways of weighing the relative importance of various factors are adequate, i.e., through human constructs such as models, lists, and meetings (e.g., the NEPA process, Cantor 1996), especially as they bring human values into the process.
• We are doing things that meet the tenets of management as best possible.
• Human intellect is sufficient to continue following conventional approaches rather than be used to find replacements for failing parts.

These beliefs may be a primary factor in preventing progress toward an approach that will replace many flawed aspects of conventional management and begin to solve the problems before us. The fact that human constructs involve varying accuracy in their depiction of a particular reality is not accounted for in conventional management. Partial information only indirectly related to management questions is often brought to bear in decision making. A major contribution made by systemic management is to accept the human limits behind these errors to move to defining appropriate choices and uses of information. Science is used in a way that observes exemplars that work (consonant realities are the models, perceived and represented directly in human constructs). This is done in a way that largely relieves us from having to shoulder the load of integrating complexity and dealing with the problems of conflict, control, and repercussions. This integration is accomplished *a priori* as shown in Figure 1.4 (and as will be treated

again in Chapter 6, see also Belgrano and Fowler 2008, Fowler and Hobbs 2002). We have seen hints regarding the most direct representations of reality possible (i.e., measure of real phenomena, empirical observations of real-world models) and how they can be used instead of models constructed through incomplete combinations of limited indirectly related human concepts. The information from real-world models has been illustrated in the preceding chapters (especially in the example for the eastern Bering Sea in previous chapters). These issues will be revisited in Chapter 6.

With this in mind, further but still superficial development of the concept of systemic management is seen in the following section where it is extended to ecosystems and the biosphere. It explores solutions to the dilemma of conflicting objectives, finding balance among risks and benefits, and reconciling the opposing forces in nature.

4.5 Including ecosystems and the biosphere in management

In conventional forms of management, we often isolate or separate the focal level of biological organization from the others (reductionism and dualism). As shown above, this approach results in irresolvable conflict and is a major inadequacy of conventional/transitive approaches: we do not know and cannot find ways to balance the objectives for individuals with those for species. The common dilemma is our inability to find ways to simultaneously meet objectives among the various levels of biological organization (from individuals to the biosphere). We have insight to the kinds of dynamics involved in hierarchical conflict through the work of people like John Nash (1950a,b). This insight does less to provide tools to figure out what to do than it does to better understand that the options for what works (and will work if we try them) appear in a natural system.

To deal with the problem we confront in finding consistency in our management at the various levels of biological organization, we must fully understand that humans—as individuals, communities, and as a species—participate in ecosystems and the biosphere. A key step forward is realizing that the same is true, of course, for all other species. All are made up of subatomic particles, atoms, molecules, elements, compounds, cells, and organs, and all occur in a larger context of ecosystems, the biosphere, and the rest of the universe. All are involved in dynamics, processes, interrelationships and complexity. Therefore, management that includes ecosystems and the biosphere should provide for the needs of individuals of all species, including humans. It should also meet species-level needs, exemplified by defining or prescribing the conduct of human affairs to account for the risks of our own extinction. It means management to include all other species so that sustainability is extended to them and their ecosystems. This means regulating human activities in relationship to ecosystems while simultaneously regulating human activities at other levels.[37] To be comprehensive (holistic), such activities would be carried out to achieve the variety of goals listed above, including opening the option for ecosystems and the biosphere to return to normal status, even if we do not know about the limits within which such qualities are expected to vary. Ecosystems and the biosphere should not be subjected to abnormal influences (or observed to exhibit abnormality) any more than should individuals or species.

Chapter 5 continues developing systemic management as it applies at various levels, including human interaction with the biosphere (Fuentes 1993, Huntley *et al.* 1991, Lubchenco *et al.* 1991, Myers 1989, Vallentyne 1993). Common among the principles and the approaches discussed at the Airlie House meetings and similar venues is the necessity of including ecosystems and the biosphere. This is a step toward considering complexity, particularly including higher levels of biological organization in the hierarchy of life. Occasionally ecosystems or the biosphere are proposed as a focal level of management. Both are levels of biological organization for which the management tenets should apply. However, in all cases of transitive management, achieving focal level management objectives transitively often creates problems at a different level. Without the intransitive (human self-control), there is no recognized process in current management philosophy for solving the resulting dilemmas and paradoxes.

As seen in the previous sections, conflict is accentuated rather than resolved if we approach

"ecosystem management" or "biosphere management" transitively, with ecosystems or the biosphere as focal units. This section considers how we can carry out intransitive management in a way that resolves conflict and considers communities, ecosystems, and the biosphere with all their inherent and intrinsic interactions, processes, and other aspects of complexity.

4.5.1 Accounting for complexity

Expanding the definition of management to include the conduct of human affairs at the species-level is a step toward including ecosystems because we are one of the participating species, and managing our species' interactions with ecosystems is a single-species application. As a species (and as individuals, families, or communities), we influence ecosystems and the biosphere just as do all species (and their components). Our roles, our influence, and the position we take within ecosystems are elements over which we have some level of control. Defining management as regulation of these factors and influences can include (but not be restricted to) the goal of avoiding abnormal risks of human extinction along with the host of other risks all taken collectively (keeping in mind that risk cannot be eliminated entirely). In progressing toward a definition of management this way, we are left with need for guidance as to how to proceed, whether as individuals or as a species, especially in finding solutions to natural conflict. For our species, this guidance is provided in information exemplified by patterns showing human abnormality (Fowler 2008). Finding a position for humans within the normal range of natural variation would account for complexity in that complexity is integrated in such information (Fig. 1.4), through what is observed naturally rather than through human constructs (i.e., moving stakeholders from the top row to the bottom row of Fig. 1.1).

4.5.2 Resolving conflict

How do we account for all of complexity, and at the same time find the necessary balance among the opposing forces that emerge as conflicts that are irresolvable in conventional (primarily transitive) approaches? We do so by managing systemically wherein reality-based patterns provide consistent guidance (Hobbs and Fowler 2008). The resolution of conflict is one of the main accomplishments achieved by following observable empirically successful examples of sustainability. The patterns we see are products (Fig. 1.4) of the very opposing forces that give rise to what we experience as conflict ending in debate (Fig. 1.1). Conflict is resolved when interpreted as the forces integrated by observed phenomena (we understand this based on Nash's insights without using the insights but rather use the empirically observed results of natural processes). As we will see in Chapter 6, empirical solutions exist in the examples provided by other species, as tenuous and vulnerable as they are. As such, the sustainability, options, and guidance exemplified by other species are made available to us in information about the limits to the natural variation among them, as seen in species-level patterns (e.g., frequency distributions, Fowler and Hobbs 2002). These species have emerged, through processes including the trial and error dynamics of selective extinction and speciation, as a trial and error (or Bayesian-like integration, Appendix 4.4) process, as exemplars—some more so than others—of the very limited sustainability that is achievable. They have so far escaped extinction by solving the problems of conflict experienced in the opposing forces among the various processes of nature. Humans potentially have an advantage over other species due to our capacity for creative imagination, intellectual reasoning, introspective epistemology, and behavioral adaptability and resilience. Guided by the experience of other species, we may be able to optimize these advantages. The alternative is to succumb to the forces of nature, and fall prey (for example) to Combinations 5 and 6, of Table 3.1—especially Combination 6—through forces that lead to the risk of our extinction.

The selective forces of nature, whether at the level of the gene, individual, or species, present all species with the same dilemmas, manifested in the conflicts and management issues we must address. Contributing to conflict are the opposing forces in natural selection generically described in Chapter 3 and modeled in Appendix 3.2. Empirically observed species frequency distributions exemplified in Chapter 2 represent the balance struck in natural systems—today of course, including all human abnormality. These balances include those among the various forces of the nonevolutionary processes and between these forces and those of natural selection at all levels. The categories represented by the most numerous species represent the best examples of a full accounting of risks as manifestation of the balance struck in nature among the same naturally occurring, opposing forces that pose dilemmas for humans. Among the risks accounted for are those that simultaneously fully account for the benefits and the interplay among them. For example, a normal risk of extinction equals a normal chance of continued existence as a species (i.e., sustainability; Management Tenet 6).

Thus, under normal circumstances, existing individuals, species, and ecosystems represent transient evolutionary solutions to the problem of long-term sustainable existence within living systems. Existing life forms are the result of a successfully achieved balance (as dynamic balances or trajectories toward elusive balance) among opposing forces, even if it is temporary in geological time scales. By mimicking the solutions existing individuals, species, and ecosystems offer by example, we find a solution to the fundamental dilemmas of management outlined above. This approach is as value-neutral[38] and objective as possible; these life forms represent examples of options that account for risk to exemplify sustainability—they are not yet extinct. The world's ecosystems and biosphere have endured human abnormalities for only a geological instant compared to the time we and most other species have been around.

How is it that we solve the problems of conventional management by adopting systemic management? What is it about Figure 1.4 that means we have dealt with complexity (and have dodged the dilemma of conflict if we use empirical information to guide management)? The following section explains how the information in Chapters 2 (and, later, 6) can guide us to achieving sustainability in a way that addresses these questions by managing our relationships with other species, ecosystems and the biosphere.

4.5.3 Nature's Monte Carlo experiments in sustainability

What do the dynamics of selective extinction and speciation have to do with patterns being integrative accounts of the infinite of reality? What does Chapter 3 have to do with Figure 1.4 in accounting for complexity? As described in Appendix 4.4, an analogy can be drawn between selective extinction and speciation on one hand, and the Monte Carlo (randomized experimental) aspect of Bayesian statistics (Howson and Urbach 1991) on the other.[39] Species are nature's trial-and-error models of success, as tenuous as it may be, reflective of all the factors involved in their emergence. These factors take the place of data used in Bayesian statistics wherein probabilities are represented in quantitative models.[40] The probability distributions represented by species frequency distributions carry information, part of it in genetic code (DNA), parallel to the computer code used in Bayesian models (Fowler 2008). Thus, species frequency distributions, as probability distributions, reflect the constraints known to operate in natural systems (Fowler and Hobbs 2002). They are cybernetic in nature. Existing species represent an integration of all factors in their environment, history, explanation, and nature (Belgrano and Fowler 2008; Fig. 1.4). They are, in part, products of natural selection, including those expressed through selective extinction and speciation—the risk of extinction is taken into account. They include an integration of all other factors that come into play, such as measurement error and the complete suite of ecological mechanics, including anthropogenic influence and the associated belief systems. The opposing forces that give rise to irresolvable human conflict in conventional approaches to management are among the factors involved. The resolution of dilemmas is inherent to natural patterns. As such, natural patterns represent practical information for use in systemic management, characterized by intransitive coping, succeeding, surviving, or getting by without abnormal influence on larger systems such as ecosystems and the biosphere—largely prevented by the constraints of the more inclusive systems.

4.5.3.1 Including other hierarchical levels

Natural selection, combined with the other forces of nature, provides a valuable source of information for guidance in conducting human affairs, precisely because of the way the resulting patterns integrate, and account for complexity (Fig. 1.4). The frequency distributions of individuals, species, and ecosystems serve not only as normative information for evaluating systems at the respective levels, but also as a basis for identifying the need for action. Action without control is ill-advised, but control can be exhibited in human action guided by examples of success. In conventional transitive management, the complete set of risks and benefits of action are rarely (if ever) identifiable, owing to complexity and lack of experience. However, solutions to such problems have evolved as exemplified by other individuals, species, and ecosystems, emerging within the context of this complexity and attendant risks (i.e., nature's multilevel Nash equilibria, Nash 1950a,b). The risks are part of the forces within the evolutionary processes contributing to the observed species frequency distributions.

At the *individual* level, we already use such an approach. For example, comparisons among individuals within a species establish the most risk-aversive body temperature, body mass, blood pressure, respiration, and ingestion rates, even our interactions with other individuals and species: values close to the average (or those that maximize biodiversity, Fowler 2008). We learn from role models; as individuals we are partly who we are based on our interactions with other people. Successful forms of interpersonal relationships emerge, just as body temperature evolved as a sustainable solution to a complex set of factors, including chemical, physiological, and environmental elements. Even if one knew all the factors involved and how they operate mechanically, it would be impossible to predict an optimal body temperature *a priori*. Likewise, when faced with managing multiple levels of biological organization we can be guided as a species by existing examples of successes represented by other species. We can understand ourselves as a species defined, in part, by our interactions with other species.

We determine safe body temperatures, heart rates, and behavioral traits by the empirical examples of other individuals. In parallel (including another hierarchical level, or logical type, Bateson 1979), other *species* represent empirical examples of sustainability that are informative for our species. For example, when faced with the need to estimate a sustainable harvest of a resource species, guidance is found in the frequency distribution of consumption rates by other species that feed on the resource in question (e.g., Fig. 2.6). Restricting our consumption rates to avoid the abnormal within the limits observed within empirical data on consumption rates is an example of intransitive management applied in a single-species approach (control of our species in interaction with another). At the *ecosystem* level, this approach would mean confining overall human consumption to the limits of consumption rates observed for other species from a particular ecosystem (e.g., Fig. 2.7, Fowler and Hobbs 2003). At the *biosphere* level, it would be achieved by confining consumption by humans to within the limits of natural variation in consumption among nonhuman species for the earth as illustrated in Figure 2.10.

Other ecosystem level applications involve, for example, the advisable numbers of species to harvest.[41] At the biosphere level we are also faced with the larger question of how many humans can be sustained. Here distributions regarding population density (Figs 2.16, 2.23, and 2.31), geographic range size (Fig. 2.27), and population size (Fig. 2.17) provide guidance.

4.5.3.2 *Effort and sacrifice (management) rather than conflict*

Existing species represent the current best examples of adaptive management as carried out through natural processes.[42] Characteristics shared by the species most numerous in their frequency distributions are likely to represent evidence of optimal forms of successful risk aversion and benefit maximization—mutually workable relationships between them and their environment. Risks and benefits include those associated with ecological mechanics and with the reciprocal interaction between hierarchical levels—between systems and their components. This includes both the supportive (e.g., provision of materials and services) and the limiting (materials and services are finite and shared among components) functions of ecosystems for humans as one of many components. No risk or benefit is excluded from the information of frequency distributions, and all are included in proportion to their actual relative importance. Examples of sustainability include the goods and

services provided/used within the limits established by forces operating within ecosystems (or the biosphere). Characteristics of common and enduring species represent minimum risk solutions while characteristics of rare species or kinds of species with high turnover represent maximum or higher risk solutions. Characteristics of highly abnormal species (completely outside the normal ranges of natural variation) represent extremely pathological and high risk situations (as often the case for humans; Fowler and Hobbs 2002).

For the human species to achieve sustainable species-level characteristics that emerge in the constraints on variation among species will, in many cases, require individual-level sacrifice. What were unresolved conflicts now become identified sacrifices in the form of actions necessary for achieving sustainability. These are sacrifices individual humans can make to contribute to sustainability in achieving balances inherent in the forces of nature or complexity in general. Costs are inherent to achieving sustainability by accounting for all risks, including risks such as our own extinction. These are short-term human costs for long-term human good (including long-term good for all associated systems—for example, other species, ecosystems, and the bioshpere).

Individual sacrifices (or, more positively, contributions) toward achieving sustainability go well beyond contributing to an environmental organization, recycling, and driving eco-friendly vehicles. The extent of changes necessary to achieve sustainability will be more clearly seen in Chapter 6 (e.g., see Fowler 2008).

As is becoming increasingly clear to many scientists, many of the problems we face are tied to overpopulation, keeping in mind the very important fact that each problem must be treated directly on its own merits. In a typically density-dependent fashion, the need for such sacrifice will be much less after (and if) the human population can be reduced or is reduced by systemic forces (e.g., global pandemic). The issue is to decide that accounting for risks collectively (including that of human extinction) is a "good" worth working for. Sustainability can be understood as a value comparable to that of the normal for individuals (e.g., body temperature, blood pressure, etc.) but extended to embrace all forms of life simultaneously. Sustainability is a "good" worth sacrificing for, while experiencing the disadvantages of the individual-level changes required to achieve desired species-level change. Such individual level contributions are a matter of making up for lost ground from a history in which individual level needs and short-term issues have been considered of greater importance than protecting ecosystems, the biosphere, and future generations of humans from current human excesses.

Management cannot be restricted transitively to the focal level of ecosystems as described above in order to achieve "ecosystem management". Lack of control, unpredictable reactions, and intensified conflict lead to feedback and consequences we want to avoid. Problems at the ecosystem level drive our consideration to even larger systems, such as the biosphere. How far can we extend this process to include larger and larger systems (or more and more inclusive systems) before realizing we are not in control? We are left with only the alternative of controlling ourselves to limit our influence on all systems. This is the core of management because individual, species, ecosystem, and biosphere cannot be separated in terms of application and complexity. There may be a transcendent quality to the emergent nature of the more inclusive systems but it always includes its components through the interconnected quality of nature.

Mimicking nature's examples of sustainability at the species level (Fowler 2008) places proper importance on ecosystems, partly because ecosystems are made up of individuals and populations of unique sets of species. Species are exposed to the constraints of larger systems such as ecosystems and the biosphere. Greater weight is given to the species level than the individual level because species are made up of individuals. Even greater weight is given to ecosystems and the biosphere for the same reason. This pattern in weighting stems from the fact that constraints of more inclusive systems confine the constituent elements. It is a natural weighting that negates the option of human design. The reverse effects, of a similar nature, emphasize that parts of systems can never be ignored (e.g., if all individuals of a particular species die the species goes extinct). Other than the balance struck in the evolution of existing species

and sets of species, there is no means of weighting things for which we have no measures. A transitive approach cannot be used to sustainably "manage" ecosystems, and even less so the biosphere.

4.6 The case for systemic management

We are left with the option of guiding human affairs, by the example of other species' successes, toward sustainable species-level participation in (influence on) ecosystems and the biosphere. This approach meets Management Tenets 4 and 5 and accounts for the various hierarchical levels of biological organization with empirical examples. It is an intransitive approach, which applies Management Tenet 3. It is based on the acceptance of individuals, species, ecosystems, and the biosphere as inseparable; this acceptance of inseparability is needed to meet all of the management tenets, especially Management Tenets 2, 3, 4, 6, and 7. It is not called "ecosystem management" partly because that term implies potential for focal level success in transitive management, an approach that has been rejected as impossible. It is also not called "ecosystem management" because it is not restricted to our interactions and influence on ecosystems (it includes other species, groups of species, and the biosphere). Systemic management considers individuals, species, ecosystems, and the biosphere as parts of an amalgamation that makes them inseparable differentiated elements of reality. Our interactions with all levels make up the parts of systemic management.

In the game of evolutionary dynamics, the objective is to stay in the game. Although the presence of any species is tenuous, the genetic code of species that have succeeded in staying in the game is information about how the game might best be played.[43] This information is illustrated by the species-level patterns shown in previous chapters (with more to be seen in Chapter 6) and especially those yet to be discovered and better portrayed in their interrelationships and relationships with the environment. Each distribution includes insight from information found in genetic codes.

Through systemic management, humans are the managers and also the ones who are managed.[44] We have insufficient control over other elements of life, and the laws of nature, to proceed otherwise and we court disaster by pretending that we can control. Although far from complete (we cannot control that our influence has consequences), our greatest control is over ourselves to find a sustainable level of influence. As argued above, we are best served in this regard by conducting human species-level affairs in accord with examples of sustainability demonstrated through the limits observed in variation among nonhuman species. One of the many reasons for doing so is to account for the risk of human extinction. Extinction is one of the processes that contribute to rarity of species within the tails of frequency distributions (Chapter 3). In following the guidance of information from species frequency distributions, the elements that have been identified as important to account for in "ecosystem management" are integrated into the process (see Appendices 4.1 and 4.3 and Management Tenets 1–9, Chapter 1 and above).

In practice, this means finding the species-level patterns consonant with management questions that we face (Belgrano and Fowler 2008). Decisions and policy can then be based on mimicking natural examples of sustainability by constraining human activity so as to avoid abnormalities. This is an implementation of Management Tenet 5 (Fowler and Hobbs 2002, Mangel *et al.* 1996). Proceeding this way is precautionary. We will see numerous examples of guiding information in Chapter 6.[45] A primary scientific constraint and difficulty is finding information that represents the normal range of natural variation in a world so heavily influenced by humans. The primary challenge, overall, is management action based on that information.

4.7 The eastern Bering Sea example

The typical transitive approach to managing an ecosystem such as the eastern Bering Sea is to manipulate the abundance of nonhuman species and thus the composition of the ecosystem. Typical goals are to stimulate greater production of resources for harvest/utilization by humans. Typical methods are to apply incentives or disincentives to limit or promote fishing of certain species, or to close certain areas or seasons to fishing. Often, in ecosystems more generally, these methods are based on knowledge of predator/prey relationships, with the thought that we can replace competing predators, or the belief that if we control predators there will be more of their prey available for human use. For example, a species such as the arrowtooth flounder in the eastern Bering Sea might be seen as over-abundant (population size beyond the normal range of natural variation for such species), as reflected in its position in the frequency distribution of consumption of walleye pollock (extreme right of the top left panel of Fig. 2.6). One option, in conventional management, would be incentives to increase harvest of the flounder to reduce its population.

Such an approach always has consequences that are unanticipated and contrary to our desires. Some experts or stakeholders involved in the decision-making process might foresee some of the consequences and argue against actions being contemplated—creating conflict without resolution. This is typical of conventional management, stakeholders have varying opinions and interpretations regarding specific parts of the information used in management (top row Fig. 1.1). It does not result in objectives consonant with management questions in a way that accounts for complexity. It does not account for secondary or other higher-order effects the actions will have, or the fact that many of these effects are both unknown and unpredictable. It does not consider evolutionary reactions to human influence that have already been set in motion (also unknown, unpredictable, and especially unprovable with current forms of science and logistic constraints), nor those that would be initiated by such action. The repercussions for us as managers are unknown. Species-level patterns that fail the test of consonance (match with the management question, Belgrano and Fowler 2008, Hobbs and Fowler 2008) are not the basis for managing other species transitively.

In the eastern Bering Sea, the process of conventional management is perhaps most clearly exemplified in the certification of the walleye pollock fishery as sustainable by the Marine Stewardship Council in 2004. This typifies action by a group of scientists, managers, and other stakeholders in rejection/violation of Management Tenet 5. The decision to certify the fishery as sustainable was an exercise of decision-making depicted in the top row of Figure 1.1 and described above. The decision rejected the abnormality of the commercial catch of this species (Fig. 1.7 and top panel of Fig. 4.1, Fowler 1999b, Fowler and Hobbs 2002, 2003, Fowler and Perez 1999) as a problem to be solved. Instead, the certification was based on artificial combination, interpretation, translation, and evaluation of less relevant piecemeal information. Also involved, of course, were the values, belief systems, thinking, and habits behind conventional management. Economic factors played a role. In this example we see a clear connection between conventional management and abnormality; certification embraced the abnormality as acceptable. In contrast, fully embracing Management Tenet 5 (especially in combination with Management Tenets 1 and 2) would lead to the conclusion that the fishery is certifiably unsustainable. Overall, this example makes clear the vulnerability of current management to the subjectivity and fallibility of human nature. As in all of conventional management, the action taken by the Marine Stewardship Council was not malicious, or ill intended. It was a matter of doing the best possible using the paradigm behind conventional thinking in attempting to achieve an optimal outcome—again defined through conventional approaches and the values involved.

The same holds true for all aspects of management in the eastern Bering Sea. The harvest of resources is carried out without objectively addressing questions such as: "What is sustainable age or size selectivity within the commercial harvest of walleye pollock (or any other species)?" For the most part, such questions are not posed as clear management questions (top row, Fig. 1.1). What portion of the eastern Bering Sea should be set aside in marine protected areas? What portion of the geographic range of any particular species (e.g., Steller sea lion, fur seal, walleye pollock, thick-billed murre) in the eastern Bering Sea should be set aside in marine protected areas? Decisions to harvest at rates, in locations, at times, and with selectivity that are abnormal compared to other species clearly contribute to the abnormal influence we have in, not only the eastern Bering Sea, but all ecosystems. Again, such decisions are not made maliciously. In fact, those making

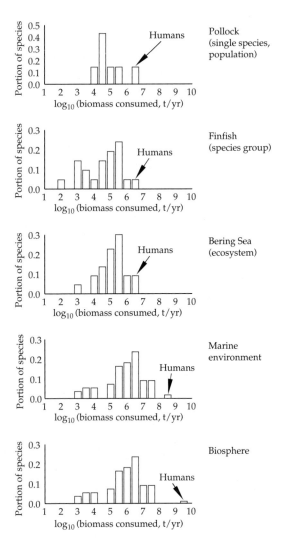

Figure 4.1 Species frequency distributions showing the rates at which marine mammals consume at various levels of biological organization within the eastern Bering Sea, the marine environment, and the biosphere showing consumption by humans in comparison to provide initial indications of the kinds of changes we will have to undertake to fall within the normal range of natural variation (from Fowler 2002, Fowler and Perez 1999).

the decisions are responsible caring people. They are people doing the best job possible within the confines of conventional management, knowing that we humans need food, the fishing industry depends on catches that are economically viable, and that the productivity of resource populations is stimulated by being harvested.

By contrast, systemic management has the goal of providing for human needs while simultaneously and consistently ensuring that such needs are sustainable and can be met sustainably. All species depend upon the sustainability of the services that ecosystems provide. The methodology involves the use of species-level patterns (species frequency distributions) as guidance for sustainably harvesting fish biomass, by mimicking other species' proven (though imperfect) examples of sustainability. For example, increased harvests of arrowtooth flounder might be part of the strategy, but harvest rates would be restricted to avoid abnormal levels (i.e., a pattern such as shown in the top panel of Figure 4.1 for walleye pollock, using such a pattern that would be developed for the arrowtooth flounder).

Such an intransitive approach is not restricted to providing guidance for the harvesting of an individual species from a particular ecosystem such as the eastern Bering Sea. It also provides guidance for harvesting from species groups and other ecosystems as well as extensions to include the entire marine environment and the biosphere (Fig. 4.1).

We are left with several issues that have yet to be adequately addressed. In particular, there is the matter of human influence as a factor in contributing to what species-level patterns are today (Fig. 1.4). The sustainable take of walleye pollock as exemplified by nonhuman consumer species might be a greater fraction of standing stock biomass if their populations were larger. However, if the predator populations were larger, the walleye pollock population likely would be smaller. These and other factors cannot be ignored in dealing with complexity (as required by Management Tenets 3 and 4). They are part of the difficulty we face as a result of a history of practicing conventional management without accounting for complexity.

However, harvesting one species is only one isolated example of management that needs adequate guidance. There are directly related questions:

• What should management establish as an appropriate age composition for the harvests?
• What should the harvest rate be if we take only adult fish?
• What are the optimal levels of harvest in terms of numbers of individual fish rather than biomass?

• How can we best allocate takes over season and space (including depth)?

Management expands to similar questions for other species, groups of species, other ecosystems and the biosphere:

• What is the optimal number of species to harvest?
• What amount of biomass and numbers are to be taken from entire ecosystems?
• At what level should we be producing pesticides or carbon dioxide that find their way into various ecosystems around the world, including the eastern Bering Sea.

As we are beginning to see, these are also issues that can be addressed through systemic management.

By ignoring complexity, conventional single-species management may create what we would conventionally evaluate as problems at the ecosystem level (including extinctions). Systemic management might do the same. There are two points: (1) there are consequences to our management no matter how we proceed and we will often judge some negatively; and (2) we cannot confine our management to one issue because complexity demands that we address as many issues as can be identified. If extinction rates are abnormal, we are responsible for doing an inventory of any ways we might be contributing to see if there are human abnormalities to be corrected.

Pollution may be a factor in the eastern Bering Sea and the problems we see in that system. As we will see in Chapter 6, application of guidance from species-level patterns would lead to restricted pest control (no other species produces as many toxins released in quantities that humans are using). We may be able to influence (rather than control) other species using guidance from species frequency distributions, but the extent of such influence will likely be quite limited compared to what we have historically attempted in the mission of control. There are benefits to an intransitive single-species management approach (e.g., we achieve sustainability, find balance in meeting our needs with the capacity for systems to meet them, and reduce the prevalence of pathologies in various systems). However, the consequences continue to involve what we will evaluate as problems and many of those will occur at other levels of biological organization. Reducing the harvest of walleye pollock causes individual fishers to lose income or jobs. Harvesting walleye pollock at any level results in competition with other predators. The harvest affects the ecosystem in which pollock occur, and some of these effects may be undesirable to us, but in all likelihood would be within the normal range of natural variation.

In management, control rests with humans as self-control, difficult as that may be. The numbers of species we choose to harvest, the amount of biomass we choose to remove from an ecosystem, and the amount of biomass we choose to harvest from any particular species can be based on limits apparent in species-level patterns. The quality of the data we have in hand for producing useful species frequency distributions is a factor yet to be addressed.

4.8 Summary and preview

This chapter has presented examples of the failure of conventional management and some of the reasons for these failures, and has pointed toward systemic management as an alternative to solve these problems.

Conventional management fails to meet the nine tenets of management. Human interests (especially factors such as economic factors) are over-emphasized in comparison to an objective accounting for our roles in ecosystems and the biosphere. That is, we fail to meet Management Tenets 1 and 2, believing that our emphasis on humans assures adequate consideration. Conventional management fails to meet Management Tenet 4 in lacking consistency in consideration of other elements of reality, and Management Tenet 3 in not accounting for reality in general. In principle we know we are finite in being human and cannot account for all unintended consequences of our actions; yet these principles are largely ignored in conventional management. All too often we believe that we have control and lead ourselves into excessive influence, frequently to find ourselves involved in abnormal interactions with the nonhuman—in violation of Management Tenets 2, 5, and 6. We believe we can generate management advice (e.g., through meetings, congresses, panels, expert advice, political agendas, special interest groups) rather than simply looking to nature for guiding information through direct observation. We proceed this way, largely ignoring Management Tenet 3 and also 2, 4, 5, 6, 7, and 9 in spite of practical experience that tells us it is humanly impossible to account for complexity and reality otherwise. Among the factors we have failed to take into account are the many risks and their interactions—all parts of reality—and have focused on sustainable development rather than achieving sustainability. The choice and use of information in today's management involves serious logical errors. Science is used to emphasize isolated factors rather than observe realities directly related to (consonant with) management questions in a way that meets all nine management tenets. As a result we are left with few clear directions, have few objective standards, lack adequate criteria for management action, and face seemingly insurmountable problems for which we are largely responsible.

These failures leave us with systemic management as an alternative defined by the tenets of management and, therefore, a way forward so as to adhere to them all. Chapter 5 describes how well systemic management is defined by, and adheres to, the tenets of management in a way that solves many of the problems so clearly plaguing conventional approaches.

CHAPTER 5

Why systemic management works

> Water is H$_2$O, hydrogen two parts, oxygen one, but there is also a third thing
> that makes it water and nobody knows what that is.
>
> —D. H. Lawrence

The preceding chapters provide the background needed to understand the foundations of systemic management—a significant part being that of avoiding the abnormal. This book focuses primarily on the species-level aspect[1] of such management—directed at avoiding human abnormality at the species level. Comparing our species to others (Fowler 2008, Fowler and Hobbs 2003, Figs 1.7, 4.1) provides a basis for evaluating our species. Understanding and avoiding the abnormal allows for setting goals, and evaluating our progress in achieving them. If there is hope of achieving sustainability, the process must involve our own self-control; one of the first steps is to understand ourselves as a species.[2] Regulating and limiting ourselves as a species must ultimately be done by individuals, within a context of incentives, disincentives, social norms, and education created by collective human institutions and effort.

However, thinking or behaving as if the human species is separable from (not a part of, not subject to the laws of) reality will not work—a lesson learned from experience in "resource management". Systemic management addresses the state of all other systems, but not as systems under our control. Rather they are understood as systems that do, and will, respond to our actions. As systems that include self-regulating and homeostatic dynamics (Brown 1995, Camazine *et al.* 2001, Heylighen 2003, Solé and Bascompte 2006), ecosystems and the biosphere are capable of recovery from abnormal human influence and will respond to management to relieve them of abnormal human influence. Our objectives for other systems will be attained indirectly through their reactions to such management, predominantly in the direction of sustainability—a state of health or normalcy. Otherwise, the combined responses of these systems to human influence will pose risks we clearly wish to avoid for all concerned. These are the risks that prevent the accumulation of species with characteristics beyond the normal ranges of natural variation as displayed graphically for patterns shown in Chapters 2 and 6; the risks that result from the homeostatic dynamics of systems such as ecosystems and the biosphere. These include risks presented by ecosystems with abnormal qualities.

Systemic management is guided by a combination of factors familiar to the recent scientific treatment of complexity,[3] but practical application of this guidance requires replacing at least some elements of current management.[4] Previous chapters described how systemic management considers ecosystems and the biosphere as biotic and abiotic environments in which species, including humans, evolve as both parts and products, and as systems in which such components exert a variety of influences. This chapter describes how systemic management does more than embrace all of the nine tenets of management laid out in Chapter 1. It does so, fully, consistently, and simultaneously so as to largely solve other problems with conventional management as described in the previous chapter. Systemic management is consonant reality-based management. Where applied, it fully accounts for the inclusive aspects of complexity,[5] and includes as one of its parts the management of our interactions and relationships with ecosystems to meet the need for more effective "ecosystem management" or "ecosystem-based management" (ecosystems are part of reality). Thus, systemic management is consistent at the individual, species, community, ecosystem, and biosphere levels,[6] and provides guidance for setting goals and making decisions. The limitations of systemic management are discussed, as is the importance of extinction as a management issue. The chapter concludes with a basic protocol for applying systemic management with examples from another visit to the eastern Bering Sea.

5.1 Systemic management adheres to the tenets of management

Haeuber and Franklin (1996) said that "…we may not be able to define [ecosystem management], but we know it when we see it". They make the case that we have made a great deal of progress in developing criteria that should be met by "ecosystem management", but we do not have a widely accepted form of management that meets these criteria. Accounting for complexity drives the need beyond ecosystems; management must apply to all levels of biological organization, including the biosphere (Fowler 2002, Fuentes 1993, Huntley *et al.* 1991, Lubchenco *et al.* 1991, Myers 1989, Vallentyne 1993) and the Earth (Clark 1989, Myers 1993). It must account for ecological mechanics, natural selection, and processes involving the flow of nutrients, predation, and competition—all parts of complexity.

This section evaluates how well systemic management meets the nine management tenets first laid out in Chapter 1 and revisited in Chapter 4 (where the failures of conventional management were documented). These tenets distill the desirable attributes of management emerging from years of effort to define it (e.g., Christensen *et al.* 1996, Fowler 2002, Holt and Talbot 1978, Mangel *et al.* 1996, McCormick 1999, and the references in Appendices 4.1 and 4.3). Systemic management is defined as the management of human interactions with other systems so as to avoid the abnormal—a definition that emerges from this history.

5.1.1 Management Tenet 1: Including humans—management must be based on an understanding of humans as part of complex biological systems

Christensen *et al.* (1996) stress, as do many others (Appendices 4.1, 4.3), that management must recognize that there is a place for humans in the grand scheme of things. It cannot be emphasized enough that the sustainability of humans as natural elements of ecosystems and the biosphere is an integral part of systemic management. Just as a person with a seven-degree fever is a natural phenomenon, the abnormality of the human species is natural and subject to the natural processes resulting from such pathologies. As parts of systems, we are constrained by the natural laws that govern all species. Our abnormality in comparison to other species subjects us to the same forces that operate homeostatically to bound such variation. Finding a normal, sustainable existence for the human species in its many dimensions, a realistic niche for humans, is a fundamental goal of systemic management. It is a means of finding a realistic way to fit into reality.

Systemic management attempts to adjust human impact so we can sustainably coexist with other species, within ecosystems and the biosphere. It aims to increase the likelihood that humans will not be excluded from the Earth's biota through the homeostatic processes of ecosystems and the biosphere. Systemic management ensures that the needs of humans are accounted for in management, in part, through the identification of needs that are unsustainable. Such management rejects the option of doing things that lead to an abnormal risk of our own extinction, but includes extinction as one of the risks we face as a species. In systemic management, humans, like other species, are considered to directly experience the laws of nature over all scales of time, space, and hierarchical complexity. Human experience is extended to direct empirical observation through which we see all species subject to, and conforming to, these laws. We can use the constructs of models to help understand, see, and measure the manifestations of these laws, but not to fully recreate reality (Pilkey and Pilkey-Jarvis 2007). No species, including humans, is capable of sufficient control to change or eliminate these laws.

We are not making up the rules; we are learning to follow them, or ignoring them to our peril.

Humans are faced with a decision: Are we going to change or are we going to let nature make changes for us? If we are going to change, what will it be? Humans create change whether or not decisions are based on systemic guidance. With goals established using information about limits to natural variation, humans shoulder the responsibility both for taking action to achieve such goals and for the consequences of failure to do so. No lawyers represent future generations of humans or other species, and few laws overtly protect them, but if there were such laws, the responsibility of today's society would be perceived differently. Opposition to change in a sustainable direction and the acceleration of change in the opposite direction are evidence that changes of the magnitude needed are unlikely and nature will take its course. Fortunately humans have the potential for collectively awakening to the imperative for change. This potential is basis for hope without which no change will occur (Fullan 1997).

As mentioned in earlier chapters, one aspect of including humans in our concept of complexity has to do with past anthropogenic influence. Not only are we abnormal in many ways now (Fowler and Hobbs 2002, 2003), but we have been for long enough to have set in motion many changes in nonhuman systems, at least some of which are important to account for in what we do now. The ripple effects[7] of our influence have long-term consequences, including coevolutionary interactions set in motion by our past and present influence on the systems of which we are a part. Many of these effects are largely irreversible[8] and systemic management offers the option of taking all precautionary actions we can to avoid the potential of these changes resulting in abnormal burdens on succeeding generations of humans and other species—including the risk of extinction.[9] Although there may be adaptability left at the ecosystem level (i.e., we are not likely to cause the extinction of all species), the opportunity systemic management presents to our species is that of following the lead of other species: adapting to the changes we have set in motion, on top of changes that are independent of our influence. Interconnectedness and complexity (Management

Tenets 3 and 4) demand this consideration of humans and systemic management makes it happen.

Thus, systemic management makes it possible to account for the effects of current and past human influence. Comparing the population of our species with that of the dozens of other species for which we have approximate estimates of population size can be misleading if we are trying to find what is sustainable in the absence of abnormal human influence (Fig. 5.1, Fowler 2008). Because the sets of species with which we can compare our species currently are parts of ecosystems that we have heavily influenced, they themselves are not necessarily normal. We are accountable for the abnormality we have created and current species-level patterns better account for that abnormality than we find possible in conventional practices.

5.1.2 Management Tenet 2: Limited control—management must recognize that control over other species and ecosystems is impossible

Systemic management is explicitly limited to control in the human sphere—human action, influence, and characteristics. It is assumed *a priori* that, like any form of management, systemic management will have consequences, intended and unintended, over which we have little or no control and which we certainly cannot prevent. Based on the principles of complexity and connectedness (Management Tenet 3), systemic management uses empirical information to guide human action such that the goals and objectives we hold for both human and nonhuman systems may be realized among such consequences. The consequences, feedback, ripple effects and other impacts of our management are accounted for, owing to the fact that such effects are part of what always happens whether the management action is what we do as a species or as individuals.[10]

Systemic management solves the problem of our tendency to think we can decide what is best for us or other systems based on human concepts, emotions, science, special interests, political agendas, and partial information or knowledge about things not consonant with any specific management question. Our attempts to piece together the

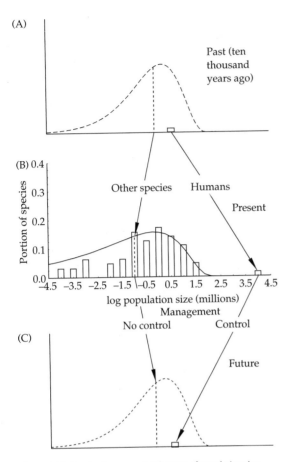

Figure 5.1 Species frequency distributions of population size for 64 species of mammals of approximately human body size, including humans, compared for a (hypothetical) past (A), a (measured) present (B) and a (potential or hypothetical) future (C) (Fowler 2005, 2008). The curves of A and C represent an approximation of the shape of the range of normal variation expected if humans were within the distribution (e.g., to maximize biodiversity, Fowler 2008), and simultaneously avoid abnormality for other measures (e.g., energy use, resource consumption, etc.). The data and curves of B show a general idea of the shape of the distribution as seen today if more of the endangered species were included. The shift in position and change in shape from (A) to (B) represents a result of the influence of humans, in part owing to the position of humans in this distribution. The potential for change between B and C is only under our control to the extent that we have control over ourselves.

metabolic rate, or range size. In systemic management, we observe and measure these things as directly as possible through scientific studies to account for complexity rather than yield to the temptation to turn simply to ourselves for the guiding information. This self-control is one way systemic management accounts for human limits.

When nonhuman systems are abnormal, systemic management makes action to repair the other systems through mitigation, engineering, or manipulation a very low priority—in many cases precludes it entirely as being beyond our control. Instead, the first order of business is to eliminate all *human* abnormality with any potential for counting among the factors contributing to observed nonhuman abnormality. Thus, we look for, and eliminate, human abnormality when problems are identified for ecosystems in the form of endangered species, shortened food chains, abnormal productivity, or elevated extinction rates.[11] Abnormality in nonhuman systems is treated as symptomatic of human abnormality until proven otherwise; abnormalities within the human realm are treated as symptoms of other kinds of human abnormalities until proven otherwise. Symptomatic relief is avoided in favor of dealing with root causes of human origin. If there is no human abnormality in all ways we can identify (particularly as a species), the occurrence of other problems would then be viewed as largely beyond our capacity to prevent. This reverses the burden of proof so that we are assured that we have done everything possible to avoid being the causative agent behind observed abnormality (human or nonhuman). Making the structure and function of nonhuman systems dependent on support and transitive action by our species is a recipe for instability not sustainability.

Thus, noting any abnormality is a stimulus to look for human abnormality of any kind. However, in the spirit of Management Tenet 3, the independent discovery of any human abnormality is cause for restorative action owing to the systemic repercussions involved (especially because so many nonhuman elements of reality are involved— some beyond knowing about and whether or not cause-and-effect links have established).

parts of reality remain a human construct rather than reality. Within reality are limits to the natural variation of such things as predation rates, CO_2 production, body size, water consumption,

5.1.3 Management Tenet 3: Complexity and interconnectedness—management approaches must account for reality in its complexity over the various scales of time, space, and biological organization

Because reality is complex (Appendix 1.1) and heavily, if not completely, interconnected, our actions have many consequences (Christensen *et al.* 1996).[12] Because of this interconnectedness, complexity is accounted for when guidance for management is based on empirical patterns and the information they provide regarding the limits to natural variation (Fig. 1.4). Such guidance accounts for various systems *and* their influences and feedback—whether those influences and repercussions are normal or abnormal. As outlined in previous chapters, the species depicted in species frequency distributions exemplify both exposure to, and influence upon, reality and its interconnectedness (Belgrano and Fowler 2008). Patterns reflect the complexity of the reality of which they are a part in its entirety. For species-level patterns, these include the extrinsic/exogenous and intrinsic/endogenous factors for each species,[13] the reciprocity among all species (direct or indirect) along with all the other elements of reality, and the consequences of that reciprocity.

Scientists cannot model this complexity, but if we could, the model would involve equations representing everything, including the reciprocity of all interactions.[14] We would evaluate any model through conventional approaches to ecosystem simulation, for example, by judging its output (behavior) against observed patterns such as the distributions shown in Chapter 2 to determine how realistic the models are; variation in the components of the model would need to be confined to the limits observed in reality. In systemic management we are simply using empirical observations of reality itself—reality as the model—rather than models we construct or concepts we develop. The real-world models chosen (and those models, as abstractions, we make to represent them) provide information consonant with various management questions. Thus, when the management question is about predation (by humans) we examine empirical data about predation, direct observation of predation, rather than simulating it. All forms of predation, by all species, are taken into account directly (along with everything else) in frequency distributions for predation rates. Therefore, we find guidance for systemic management that fully accounts for complexity when we use information on the limits to natural variation ("fully accounts for" is the infinite of Fig. 1.4).

Accounting for complexity in guiding information is one of the biggest steps in implementing systemic management (compared to conventional approaches), but this is only one step in meeting this challenge accomplished in systemic management. Such information accounts for complexity only in providing guidance for individual questions (Belgrano and Fowler 2008). However, there are many management questions. Each management question can be both refined and expanded. Each specific management question (e.g., sustainable consumption of biomass from a specific resource species, or allocation of consumption over alternative species) is a focused component to the carrying out of systemic management.[15]

In a bit more detail, there are four components to dealing with complexity as accomplished in systemic management (Belgrano and Fowler 2008, Fowler and Hobbs 2002, Fowler *et al.* 1999). Systemic management

- is based on guiding information that integrates all factors (Fig. 1.4),
- addresses the diversity of questions we can ask,
- takes advantage of known correlative relationships, and
- accounts for the complexity of interconnectedness.

5.1.3.1 Systemic management uses guiding information that integrates all factors

The use of empirical patterns incorporates a complete accounting (Plate 5.1) for complexity, owing to the fact that the patterns we see are a product of every contributing factor whether science has been able to show the relationship or not. As emergent from complexity, the patterns observed in frequency distributions provide us with an objective weighting of the importance of each contributing factor—all of the interactions and synergistic effects of the processes and elements involved. Thus, the infinite of the right side of the "equation" in Figure 1.4 is all inclusive. It includes everything from subatomic particles to galaxy clusters and all levels of biological organization and biotic systems. It includes all physical and chemical forces (magnetism, adhesion, surface tensions, gravity), and all biological phenomena and processes (e.g., evolution, extinction, mortality, predation, competition, camouflage, behavior, reproduction, embryological development, photosynthesis, coevolution, mimicry, vision, and functional responses). It includes everything in its corresponding temporal and spatial scale with all the effects of history. Nothing is excluded; everything is taken into account in direct proportion to its relative and actual/true importance. This is in stark contrast to the limited lists, models, meetings, and stakeholder conversion of nonconsonant information used in conventional management. The difference is depicted in the comparison of the top and bottom rows of Figure 1.1.

5.1.3.2 Systemic management addresses the diversity of questions

Fully implemented systemic management accounts for complexity by being applied to all management questions.[16] However, there are finite limits to our ability to think of all management questions (Appendix 5.1). Human limitations are converted from being a source of mistakes (top row of Fig. 1.1) to limits realized in being unable to identify all dimensions over which patterns can be observed. Our reductionistic limitations cannot be avoided. Precaution (NRC 1999) is necessary to account for this human limitation, and humility (Ehrenfeld 1981) is necessary to accept that we can never overcome it. Precaution in systemic management is achieved in taking all human abnormality seriously. However there is a difference between being limited in our ability to ask all relevant management questions and the limits in our ability to account for complexity in dealing with any one now recognized. Systemic management circumvents the latter. Listing questions is a problem for any form of management, but the scope of options is completely open in systemic management. By contrast, conventional management is stymied by questions such as "What is an appropriate level for the human population, consistent with sustainable resource consumption, CO_2 production, water consumption, and range size?" Thus, in spite of the advantages of systemic management, we remain

limited in our ability to bring to light all relevant questions or define all management issues. It can address many *known* management questions, fully accounting for complexity in each individual case. Systemic management is not simply a matter of achieving sustainable resource consumption; and yet, without sustainable resource consumption, we fail to completely apply systemic management.

Thus, reducing our population (Fig. 5.1, Fowler 2005) must be included with reducing the amount of energy and resources we consume, along with the CO_2 we produce, and space we occupy—all must be part of management and are in systemic management. Systemic management can also be applied to finding sustainability in the number of species that we consume, and the allocation of our consumption from alternative resource species, and over space and season (Fowler 1999a, Fowler and Crawford 2004). There is resolution of the numerous conflicts identified in Chapter 4 in the sense of achieving Nash equilibria[3] in the conflicting forces among the various levels of biological organization. Further consistency (Hobbs and Fowler 2008) results from the connectedness involved in using empirical examples of sustainability as a source of guiding information. For example, a reduction in our population (whether willfully or as a product of systemic homeostatic reorganization), will result in a reduction in things like CO_2 production, energy, and resource consumption. Likewise, reducing our CO_2 production, consumption of resources, geographic range, and consumption of energy would result in reducing our population. There is reciprocity. All would be guided by information on the limits to natural variation as exemplified in Chapter 2 (with more to be seen in Chapter 6).

One way systemic management accounts for complexity is through achieving sustainability in our interactions with the various levels of biological organization. For example, the hierarchical nature of life is involved in regulating our harvest of resources from other species, sets of species (e.g., communities), ecosystems, and the biosphere.[6] Accounting for complexity requires that we actually reduce such harvests, and then regulate them at sustainable levels.

Although we will never know how to ask all management questions, we can see the ways to ask many more. Concern about our use of water (Fowler 2008, Vörörsmarty *et al.* 2000) and nitrogen (Vitousek *et al.* 1981) can be extended to any other element or compound. Systemic management involves posing the related management questions so that matching (consonant) measures of both human and nonhuman species can be made. This allows for the comparison of human with nonhuman species to identify abnormality. Management would aim to relieve the nonhuman of any abnormal human influence.

5.1.3.3 Systemic management takes advantage of known correlative relationships

Management is called for, as part of systemic management, if the human species is abnormal in relation to the limits in variation for any particular species-level metric when comparisons involve *all* nonhuman species (e.g., exemplified by the primary production humans appropriate, Fowler 2008). If our abnormality involves excesses, reductions are required; if it involves deficiency, increases are required. Steps to alleviate such abnormality get us started in the right direction to initiate the change needed to achieve sustainability. The extent of abnormality also gives us some initial idea of the magnitude of change ultimately required. However, precision regarding the goal or endpoint (Management Tenet 9) is missing. Humans may fail to maximize sustainability at certain points within the normal range of natural variation when such variation is that observed among *all* species. Not all species are like humans in other ways. Systemic management allows us to refine goals to better achieve sustainability through overt or direct consideration of human species-level characteristics (e.g., body size as one among many).

This is achieved with correlative information (e.g., Fowler 2005). Accounting directly for correlative pattern is an essential contribution of systemic management, and demanded in cases where there are correlative relationships in macroecological patterns (such as the case of population density in relation to body size, Fig. 2.31). Human population densities comparable to those of bacteria would be intolerable (a million billion per km^2, Peters 1983[17]), even to people who are most convinced our species has no overpopulation problem. In systemic thinking this problem is solved by carefully selecting the

species in the pattern we use for comparison/guidance. When there are correlative elements to the species-level pattern, we need information for species that are similar to humans in regard to things we are incapable of changing (at least in the short term), such as our body size or metabolic rate, or in regard to things we decide we do not want to change, such as choosing to take adult fish, rather than juveniles, in commercial fishing. In this latter case, the guiding information for fishing would be based on information from predator species of body size similar to that of humans, and that make adult fish at least part of their diet. The information consonant with the management question ("What is a sustainable harvest of adult fish"—to be refined for particular species, season, and other relevant factors) is measures of their consumption rates of adult fish. A further refinement, using mammals of our body size, would likely improve the quality of guiding information. Figure 5.2 shows body size, mammalian taxonomy, and trophic level as they are directly accounted for in comparisons involving population density. Correlative information involving mammalian taxonomy and body size was used in Figure 4.1 in regard to consumption from the biosphere.[18]

Again, we do not know about all correlative relationships and must accept our human limitations in this regard as basis for precaution as well as encouragement to conduct research that will add to the information useful to the guidance of management (thus, there is basis for emphasizing further development of the field of macroecology beyond the academic motivation expressed by Lawton 1999). Systemic management, therefore, includes applied macroecology with informative patterns at the species level of biological organization.

Systemic management also directly accounts for environmental factors. Thus, another form of correlative information used to account for complexity comes from consideration of contextual factors in addition to human features such as body size. In this case, management questions are refined by addressing variable environmental factors through correlated changes in observed limits to natural variation. Predation rates serve as another example. We know that predation rates change with the density of prey (Holling 1959). Through systemic management any question regarding harvest rates

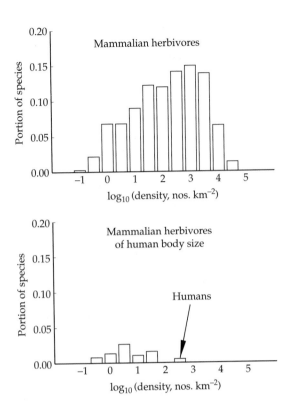

Figure 5.2 Frequency distribution for the population density of 368 species of herbivorous mammals other than humans (Damuth 1987). The top panel shows all 368 species, while the bottom panel shows the population density of humans (assuming 270 per k[2]) compared to that of 29 species of similar body size to that of humans (68 kg ± 50%), using the same scales to emphasize the size and relative location of the sub-sample that pertains to the management question relevant to humans. Note that, in comparison to mammals in general, humans appear more sustainable than is the case when the question is refined to information that more directly applies. Further refinement would include trophic level by using species of a trophic level comparable to that of humans (i.e., species with both body size and trophic level similar to humans, Fowler 2005).

accounts for the density of the resource species through such correlation.[19] Here we are dealing with biotic contextual factors.[20]

Systemic management also accounts for changes in the abiotic environment over both time and space. Management questions are refined to account for such factors the same way as for biotic factors. Here, we use correlative relationships in empirical data to account for climate, weather, season, insolation, soil types, or water quality. Thus, systemic

management takes into account both temporal changes and spatial heterogeneity. Temporal factors include such things as decadal cycles (Francis and Hare 1994), regime shifts, global warming, and glaciation. Spatial heterogeneity would include factors such as altitude, rainfall, relative humidity, and temperature. As with biotic factors, both interpolation and extrapolation would be used in taking advantage of correlations among species frequency distributions and abiotic factors over both space and time to guide systemic management.

In all cases, the factors we see as driving forces in correlative information are automatically taken into account in the complexity behind natural variation across all species, over broad time frames, and widely dispersed locations. We use correlative information in the refinement of management questions to more accurately find measurements of the set of species that best serve as role models in their response to current environmental conditions. This refinement leads to better defining the region of observed natural variation that is appropriate for our species under current conditions and in the locale under consideration. As laid out in Chapter 2, correlative relationships can be multidimensional, such that more than a few species-level characteristics can be considered simultaneously.

5.1.3.4 Systemic management accounts for complexity through interconnectedness

Because of the reciprocal interrelationships among things, avoiding the abnormal in one thing (dimension or species-level metric) automatically relieves other systems of the effects of that abnormality. Thus, reducing our CO_2 output to mimic that of other species (Fowler 2008) would address the identified problems of global warming (insofar as we contribute), oceanic acidification (insofar as we contribute), and all associated ramifications (e.g., changes in the distributions of species, disruption of embryological and morphological development, altered weather patterns, etc.) whether directly or indirectly related. We know of a few examples to list; science has yet to provide us with a full list (ultimately impossible, given the complexity involved). All such relationships and elements of complexity are involved—in the ways they are actually involved (not how we might imagine them to be involved). The reciprocity of interactions that pose a risk of extinction for our species is involved.

Even more direct interconnectedness is involved. Reducing our species' production of CO_2, for example, would have a direct impact on our consumption of carbon-based energy. Reducing our species' consumption of carbon-based energy would have a direct impact on our production of CO_2. The two go hand in hand. A reduction in our population would result in a reduction of energy consumption and CO_2 production. A reduction in any of the three would result in the reduction of our consumption of biomass (biotic resources). A reduction in our geographic range size and population density would result in a reduction in our population. These relationships may seem obvious (and to some, trivial) but they are among the realities of interconnectedness that are part of systemic management (Hobbs and Fowler 2008). These intraspecific relationships, and those involving interspecific interconnections, count among the interconnectedness accounted for in systemic management. They involve all of the dimensions over which we can measure species and involve the diversity of patterns displayed in Chapter 2 (plus those we have yet to discover).

Thus, not only does systemic management deal with the diversity/complexity of management options before us, as covered above, it also accounts for the systemic effects of our abnormalities by relieving nonhuman as well as human systems of such effects. This involves, then, both the breadth of dimensions over which we can compare ourselves to other species and the multitude of interactions and relationships that characterize complex systems.

5.1.4 Management Tenet 4: Simultaneous consistency—management must be applied consistently among its applications and must apply simultaneously at the various levels of biological organization

The search for "ecosystem management" is doomed to failure if the error-prone aspects of current forms of single-species management (identified in Chapter 4) are carried forward to higher levels of biological organization. In other words, if we attempt to conduct ecosystem management while excluding individuals and species as well as more inclusive systems (e.g., the biosphere), we will eventually learn that ecosystem management is not enough either. Similarly, management will fail if it is not based on information that accounts for finer scales of resolution, the components of individuals, cells, and molecules. Systemic management applies to, and accounts for, all levels, limited only by the human capacity for identifying management questions and capacity to conduct the research to characterize consonant patterns that provide answers. Information to guide interactions between our species and other species, between our species and other groups of species, between our species and ecosystems, and between our species and the biosphere was exemplified in Chapter 4, where Figure 4.1 shows information for application in regard to resource consumption (exposes human abnormality and indications of the extent of change needed to alleviate systems of that abnormality). Such information exists for dealing with CO_2 production, population size, population density, other forms of resource consumption, energy consumption (Fowler and Hobbs 2002, 2003), and size selectivity in resource consumption (Etnier and Fowler 2005). In Chapter 6, we will see both these and other examples of such information. All are interconnected in various ways and management regarding one will always be consistent with management for the others (Hobbs and Fowler 2008).

Observed limits to natural variation amalgamate time, spatial, and organizational scales (Fig. 1.4) to account for opposing forces among the various levels of biological organization. Observed patterns follow temporal and spatial variation; internal variation involves finer scale change. Being a sustainable species is achievable through guidance provided by species-level patterns because extant species occur within, are products of, are exposed to, have influences on, and are emergent from, complexity in the milieu of all such scales. Such patterns account for all of complexity, dynamics, and interdisciplinary issues, without our knowing the details, facts, or other explanatory information. Scale and context were identified as important to management by Christensen *et al.* (1996), not only in space and time, but also across the various levels of biological organization.

Systemic management accounts for the hierarchical nature of reality in two ways. First, guidance is based on the elements of reality that involve an integration of complexity that includes hierarchical structure. Second, systemic management proceeds in regulation of our interactions with those elements, particularly each level in the hierarchy of biological organization.[21] Consistency is achieved in the process because the guiding information is from systems governed by the laws of nature—laws that, by their nature, cannot be broken. There is consistency within and among these systems even though there are innumerable opposing forces involved in ways that seem to present unresolvable conflict. Thus, guidance for the harvest of an individual resource species will be consistent with guidance for the take from an ecosystem or the biosphere. Guidance for biomass consumption from the biosphere will limit the number of species and the geographic space over which harvests can be taken (as will guidance directly related to questions regarding the number of species taken, and geographic range), and this will be consistent with harvest from any one individual species. The same applies for other issues such as CO_2 production, size selectivity, or energy consumption, whether in an ecosystem or the biosphere.

Consistency is achieved, in part, because of the interconnectedness of things as they occur in reality (Management Tenet 3). For example, some scientists might think our CO_2 production is less important than our appropriation of net primary production (Fowler 2008). If action were taken to reduce our production of CO_2 by five orders of magnitude, it could not be done without a consistent effect on our use of energy (and visa versa). We might not complete the change needed to achieve optimal energy use, but the change would be consistent. Likewise, if the human population were three orders of magnitude smaller than it is now, both our production of CO_2 and use of energy would also be smaller. Reducing our range size might seem unimportant. If action were taken to reduce our range size, it would result in a consistent reduction in numbers of species that we use as resources and the rate at which we introduce species to ecosystems they do not normally inhabit. Such consistency cannot be

escaped, and by solving one problem others are solved simultaneously.

5.1.5 Management Tenet 5: Avoiding the abnormal—management must undertake to ensure that processes, relationships, individuals, species, and ecosystems are within (or will return to) their normal range of natural variation

This management criterion specifies that both the components of systems, and systems, should not exhibit abnormal qualities, characteristics, or interrelationships (Christensen *et al.* 1996, Ecosystem Principles Advisory Panel 1998, Fowler and Hobbs 2002, Mangel *et al.* 1996, McCormick 1999). The principle behind this tenet is that everything is limited by virtue of its finite nature. Thus we see limits to the variation of body temperature, body weight, population size, global primary production, resource consumption, and even variation in population numbers (i.e., the variability of variance has it own limits). The results of future studies will add to this list—a list barely initiated in Chapter 2. One of the basic goals of systemic management is to do what is necessary and possible[22] to ensure that processes, relationships, individuals, species, and ecosystems, are within (or will return to) their respective normal ranges of natural variation.

However, systemic management does not attempt to control the nonhuman as a way of guaranteeing that abnormal situations are avoided. Systemic management adheres simultaneously to Management Tenets 5 and 2 to consistently fulfil the requirements of Management Tenet 4. This is done through avoiding the abnormal, ensuring that it is the *human* that is controlled and within the normal range of natural variation so that normal conditions can be achieved by the nonhuman, primarily through their own processes.

As seen above, avoiding the abnormal is inherently a matter of accounting for complexity (Management Tenet 3). One way of understanding this involves understanding the role models of nature as the results of natural adaptive management experiments (Fowler 2008). Adaptive management is one way of dealing with uncertainty,

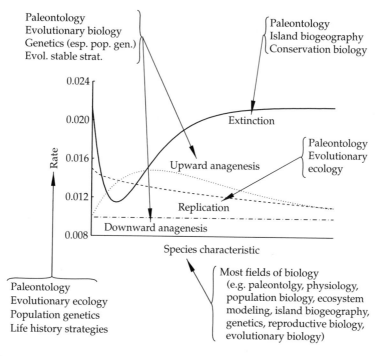

Figure 5.4 A graphic representation of the processes of evolution and selective extinction and speciation (similar to graphs in Chapter 3, esp. Appendix 3.2) showing some of the biological disciplines that contribute to knowledge of the species-level characteristics over which these processes operate and some of the processes involved. To varying extents, these disciplines also contribute limited information concerning the specific nature of selectivity and biases in the processes. Ecological mechanics are not included in the graph *per se*, but are represented by biology and all of its sub-disciplines. The lines within the graph exemplify selective extinction and speciation and evolutionary processes. The arrows indicate the limited contributions of some of the existing disciplines to our incomplete knowledge of the various processes.

exhaustively study this complexity in scientific studies. The consilience sought by Wilson (1998) is an impossibility as a human endeavor (Berry 2000) but is achieved in nature (i.e., reality is made up of all parts, including those not subject to scientific observation, Fig. 1.4).

Thus, all the academic fields combined do not cover the complexity of reality in its entirety. Even if they did, we, as stakeholders in our current position in Figure 1.1, would not be able to combine (recombine) this information in a way that accounts for everything in proportion to its relative importance. To deal with this Humpty Dumpty syndrome (Fowler 2003, Fowler and Hobbs 2002, Nixon and Kremer 1977) we take advantage of the factors brought into account by species-level patterns as natural phenomena that we observe (with the consilience found in reality, Fig. 1.4). Then science

becomes critical for what it *can* do very well: objectively, and reductionistically study a phenomenon to obtain information about its natural limits. Each part we subject to study, however, is always within its full context (Fig. 1.4). The best science for management then becomes that aspect of science that studies the pattern that is consonant with the management question being asked (bottom row of Fig. 1.1; Belgrano and Fowler 2008, Hobbs and Fowler 2008). As outlined above, mimicry based on patterns automatically includes immeasurably more than what we now know of the factors that are involved; objective, carefully conducted science is an essential tool for discovering, characterizing, and analyzing such patterns. Thus, as introduced in Chapter 4, accounting for complexity is accomplished for us, not so much by the science involved as through the natural processes that work in ways analogous to the process of Bayesian integration,[30] including the combination of processes involved in selective extinction, speciation, and evolution as it applies to change within species. Our own understanding of these factors will grow through the future development of new and existing fields of science. However, species-level patterns—past, present, and future—account for things about which we know nothing (Fig. 1.4), and fields of science we do not know to include in diagrams such as Figure 5.4 which is itself confined to our embryonic understanding of selective extinction and speciation (Chapter 3).

In the past, the field of evolutionary biology has not had as much influence on management as has our understanding of ecological mechanics; evolutionarily enlightenment is needed in our management process (Brown and Parman 1993, Conover and Munch 2002, Fenberg and Roy 2008, Law *et al.* 1993, Stokes and Law 2000, Swain *et al.* 2007, Thompson 2005). Most of the maximum sustainable yield approaches and their derivatives are based on population dynamics. The evolutionary effect of harvesting is very rarely, if ever, taken in to account; it is often assumed to be unimportant and left to be proven important through future study.[31] Conservation biology now pays attention to the risk of extinction, but management to reduce the anthropogenic causes of extinction and to account for the risk of our own extinction has not yet emerged. The things studied in the fields of paleontology and evolutionary biology are brought into systemic management, which identifies a place for humans within the limits to natural variation (Fowler 2008). Part of what species are exposed to in the diffusion-like processes of natural selection is the coevolutionary dynamics neglected in most current approaches to management.

5.1.8.1 Empiricism vs. amalgamation of science

As outlined earlier, it is not the disciplines of science *per se*, as much as the objects of their study that are important in systemic management, especially as seen in their combination.[32] What we have been doing in conventional attempts to convert the products of science to action has not prevented (and probably contributed to) the problems we face (e.g., MEA 2005a,b, Appendix 4.2). Some of this has to do with science (Appendix 5.1), but a great deal has to do with our belief that we can effectively combine the products of science rather than make an effective choice of those products without having to convert, combine, or interpret. Complex systems science and whole system thinking are behind the concept that we can treat patterns, including species-level patterns, as emergent natural phenomena.[33] They emerge through a much broader integration of the factors, processes, mechanisms, and physical manifestations of things which science can investigate but cannot adequately represent, especially in their combinations.

Rather than rely on any particular science for advice, therefore, systemic management depends on simple empirical observations—observations consonant with management questions. The advice is based on the observed not the science. Each observation is a piece of information generated by a specific kind of science, specialized in making such observations, often of things we cannot observe with our ordinary senses. Species-level patterns containing such observations integrate all such information (Fig. 1.4, with the "equation" integrating the infinite being specific to each observation). The information content of the genetic code that contributes to what species are (and thus their locations in species frequency distributions) is made available to us through these distributions. This is a major step forward because, even though we cannot have the detailed information in hand, such information is taken into account.[34] Thus, most of the unknowable[35] of such information is not ignored and although we cannot reveal the details, the unknowable is accounted for automatically. All species, and the sustainability they represent in species-level patterns, exist within, emerge from, and reflect all of the complex systems we need to take into account in meeting Management Tenets 3, 7, and 8.

5.1.8.2 Monitoring

The role of science goes beyond the process of gathering data regarding measures of natural variation and its limits—observing patterns. Monitoring, analysis, and assessment should always precede and accompany our use of resources (Holt and Talbot 1978, Mangel *et al.* 1996), our production of waste(s), or changes in the distribution of our population (or any other measure subject to management). We need to be clear about correlative relationships within species-level patterns and monitor factors known to be involved. These include the variety of influential biotic and abiotic factors in order to make direct use of such information. In the management of our use of resources, these would include the population levels of resource species and changes in the abiotic environment in which they occur. Thus, in systemic management, the population levels of resource species would be monitored to enable use of correlative information regarding predation rates that change in functional response to resource population levels. Regulation of harvests can then be based on patterns in functional response relationships. Environmental conditions are also important to monitor so that we can use species-level patterns in their relationship to circumstances in the abiotic environment. Correlative information is crucial to systemic management to account for climate and geographic variation.

As many species as possible would be monitored, both for assessing related changes in species-level patterns in response to management and for determining our relative position in the ranges of variability observed. If humans undertake change to mimic other species, the other species will react, leading to less abnormality among systems relieved of abnormal human influence—sustainably.

Monitoring the human is also an important component of information needed to evaluate progress in systemic management. The results of continuous monitoring would be made available and reviewed as with any form of management to ensure, for example, that the established sustainable harvest rates are not exceeded and that desired results are being achieved in the responses among nonhuman systems. This would apply to any aspect of systemic management. We would monitor the size of the human population, water consumption, production of CO_2, energy use, numbers of species we consume, and as many other species-level measurements as possible. An annual "state of the species" assessment for humans would serve to help assess and guide progress—always carried out through comparisons with other species. Public awareness through such a program would be of immeasurable value in motivating action.

5.1.8.3 Models as tools for research—carefully chosen for guidance

Models are useful tools and need to be included in our tool kit (Christensen *et al.* 1996), especially for research in which measurement is a primary element. They are invaluable in helping us generate hypotheses to be examined empirically. However, simulation models such as ecosystem models are an incomplete representation of ecological mechanics; they include very little regarding other factors. Left out, either entirely or in large part, are the elements of evolution and the driving factors behind evolution and ecological mechanics—factors both intrinsic and extrinsic to the system. They include little or nothing involving selective extinction and speciation. Other models such as those found in the Appendices of Chapter 3 are also ecosystem models, and are as equally valid as any other for the species sets represented by populations in ecosystems. Both the models based on ecological mechanics and those based on the mechanics of evolution, speciation, and extinction focus largely on the respective mechanics to the relative neglect of many other elements of complexity. In the end, what we observe in nature is the result of the combined set of processes, only a few of which can be partially captured in models. Models help us think about, but only superficially represent, reality (Pilkey and Pilkey-Jarvis 2007). They carry the limitation of reductionism. As with the case of natural patterns themselves, such reductionism has its value when models are carefully selected. For systemic management, their power to represent patterns as subsets of reality makes them valuable when, as models, they are consonant with specific management questions.

Thus, one form of modeling is to be encouraged. Models that mimic the "shapes" of species sets such as that shown in Chapter 2 (see, esp., Fig. 2.34) are very useful especially if variance about central tendencies is included. Such empirical models demonstrate the relationships among species-level measures; those already published provide a good start (e.g., Brown 1995, Charnov 1993, Gaston and Blackburn 2000, Peters 1983,

Rosenzweig 1995).[36] Through these relationships we better understand the end results of the dynamics we explore with the more process-or mechanics-oriented simulation models (i.e., simulation of dynamics over time). Where humans fall in the multidimensional space of such relationships then becomes more clear and the degree to which we exhibit abnormal tendencies can be assessed and used to set goals for change (management). These lead to taking advantage of correlative information. They fall in the category of what is known as statistical models (Pilkey and Pilkey-Jarvis 2007).

5.1.8.4 *Importance of social sciences*
Radical changes in individual thinking and human behavior will be needed to achieve human sustainability. The social sciences are valuable in recognizing that the complexity of human affairs is not to be ignored any more than that of ecosystems, other species, or individuals of any species. Predicting, monitoring, and assessing the sociological effects of management actions (Mangel *et al.* 1996) requires considering the opposing forces involved and the related conflicts we experience (Chapter 4) in management. More important, however, are the changes required in human systems to achieve sustainability. This will involve such disciplines as economics, religion, sociology, ethics, psychology, politics, international affairs, race relations, and epistemology.[37] Each field of study involves the investigation, documentation, and understanding of human realities. These human realities include their dynamics, processes, mechanics, and interrelationships. We have to confront our own complexity, and realize that these sciences are also limited, and the objects of their study are subject to the laws of nature along with everything else. Within the products of these fields of study is information about how change can be achieved systemically. Systemic management involves systemic human change.

It remains to be seen whether or not we humans (both individually and collectively) have sufficient self-awareness and self-control to rid ourselves of abnormality within the various species-level patterns, before nature accomplishes the task for us. The various forms of science contribute to perception, but perception does not guarantee change. The laws of nature do not change just because we perceive them, and our perception of needed change does not ensure we will make the change. Until we have a sense that we have any degree of control over ourselves, it is nonsense to think we have control over the more inclusive systems with all of the other species, their comparable complexities, and many interactions.

5.1.9 Management tenet 9: Goals and objectives—management must have clearly defined, measurable goals and objectives

Management of any kind is ineffective without a sense of which way, and how far, to go. For management to succeed, measurable goals and objectives are necessary, with special emphasis on sustainability (Christensen *et al.* 1996). This book helps construct the foundations for successful management by outlining measures (i.e., metrics) of importance to our species and its sustainability. It also points the way toward discovery of many more such measures. As pointed out by Fowler (2003, 2008) and Fowler and Hobbs (2002, 2003) (and as will be seen in Chapter 6), we have some initial benchmarks for factors important to sustainability—each of which provides measures of problems, including overpopulation, excessive energy use, abnormal CO_2 production, the consumption of too many species, and pathological resource consumption. Inherent in these species-level assessments of problems are the goals or objectives to be achieved; a measure of overpopulation automatically suggests a goal.

While ultimate goals are contained in measures made in comparisons with other species in the absence of abnormal human influence, initial goals for a variety of species-level measures can be generated.[38] Interim objectives are determined by such factors as the momentum of change in the opposite direction, political will, and extent of species self-awareness among the population. Although the broader goal of sustainability may never be precisely measurable, it includes reducing the abnormal in risks of our own extinction. The magnitude of risk requires action despite uncertainty about details; we have enough information to know the direction we should be headed and initial estimates of the magnitude of change that is necessary. Within the mix of appropriately chosen species, we are then faced with the challenge of further change in response to environmental variability.

Thus, continued refinement of goals includes accounting for past abnormal anthropogenic influence and waiting to see how other species, ecosystems, and the biosphere respond, not only to change by humans but to that of the Earth (e.g., glaciation, global warming). We need to find correlative information to help focus on the portion of the range of natural variation among nonhuman species that best applies to us humans. Also, we need to account for the nature of the biotic systems with which we are interacting, and the present and future conditions under which we are contemplating management. Further information of this kind will result in greater clarity by focusing on smaller portions of the overall range of variation, and narrowing the range of options, so that goals and objectives can become more specific. We cannot expect the optimal ranges to remain fixed. Variation is itself one of the measures of things that cannot be zero (Figs 2.18–2.20). Circumstances change and part of systemic management is to account for such variation by using correlative information regarding environmental circumstances.[39]

5.2 Limitations of systemic management

Systemic management needs to be adopted with a full understanding of what it is. This means understanding both the problems it solves as well as its own limitations. The limitations are largely human limitations as addressed in this section, and include:

- Priorities are not clearly specified.
- Species frequency distributions are inadequately developed.
- Implementation requires addressing component questions.
- Acquiring information is logistically difficult.
- Statistical and biodiversity measures are reference points—not magical numbers.

5.2.1 Priorities are not clearly specified

Systemic management has limited effectiveness in setting priorities, answering such questions as: If we are overharvesting two species of fish, as two problems to solve, which should be solved first? Is it more important to produce less CO_2 or reduce our appropriation of net primary production (Fowler 2008)? Is it more important to reduce our population or reduce our geographic range size (Fowler 2005)?

The answers to such questions cannot be known with a great deal of certainty, especially if objectivity remains a goal. For an individual management question, the guiding information within the consonant pattern accounts for the relative importance of the various factors we want to take into account in each case. Furthermore, we can now address a wide variety of questions but cannot ask all questions. However, priorities have not been specified for those we can ask.

Several guidelines seem appropriate and are presented below. It must be recognized that these guidelines may be challenged on several grounds but seem reasonable while experience leads to the development of better options.

First, priority may be given to management that contributes to the solution of a number of known problems simultaneously. Interconnectedness guarantees that this will happen in any case; even unknown problems are included. However, we know problems such as excessive CO_2 production, water consumption, energy consumption, geographic range size, biomass consumption, and overpopulation are all interrelated. Thus, placing a high priority on reducing human abnormality for any one is justified because it will contribute to reducing the abnormality in the others (as well as problems indirectly associated with any one).[40] But this does not justify avoiding attention to other problems in their own right. It acknowledges the fact that addressing the related suite of problems, without solving the population problem, will result in forced population reduction without regard to consequences for individuals (e.g., mass starvation).

Second, priority may be proportional to the hierarchical level of organization of the system we are influencing (higher for higher levels of biological organization or the more inclusive systems of which we, as a species, are a part). This seems rational on the grounds that inclusive systems control and limit component systems—their subsystems—more than the reverse (Ahl and Allen 1996, Campbell 1974, O'Neill *et al.* 1986, Wilber 1995). For example, this would mean that systemically managing our total harvest of biomass from an ecosystem would take priority over managing our harvest from any one species, and managing our consumption of biomass from the biosphere would take priority over managing either the harvest of an individual species or that from an ecosystem. The pathologies in Figure 4.1 would be treated with highest priority given to the bottom panel and lowest to that of the top panel. Certainly, full systemic management requires undertaking management at all levels, starting with issues where we have information (and promoting scientific efforts to produce more, especially where it is lacking). In any case, we would strive to avoid the abnormal in total harvests from ecosystems, even if we do not have information on the limits to natural variation in the rates of predation on a particular resource species. Given a choice, it is probably better to place a higher priority on avoiding abnormal interaction with ecosystems than with individual species.

Part of the argument for considering ecosystem applications of higher priority than single species applications has to do with the fact that, at the ecosystem level, we are automatically (but only partially) considering the effects of our influence on all the species involved rather than a particular species. In the case of an endangered species, for example, we can (in conventional thinking) ask: "How much of the production of a particular resource species should be left so that it can be consumed by the endangered species?" First, we recognize that we have no control over which species get the production left after our harvests. Then we recognize that we *can* address the question of how much would be left for *all* other species that consume the resource species in question. The management question here, as a systemic question, is: "How much of the production of a particular resource species should be left for *all* other species in the ecosystem of the endangered species?" This

exemplifies a very real ecosystem-based management question—serious questions rarely posed in today's management.

For example, Figure 5.5 addresses this question in regard to walleye pollock production in the eastern Bering Sea as an area within the geographic range of the Steller sea lion, an endangered species as of the late 1990s. We cannot control what portion of walleye pollock production that is left unconsumed by fisheries will be consumed by sea lions, or whether it will result in a suitable mix of species for an adequate sea lion diet. We can, however, address the question before us in a way that guides management action. With empirical information

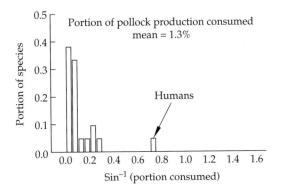

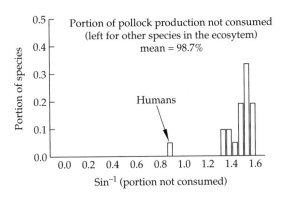

Figure 5.5 Species frequency distribution for the portion of walleye pollock production consumed (top panel) and left for the other species in the ecosystem (bottom panel), in arcsine scale and assuming a production rate for walleye pollock of 30% per year. Humans consumed about 40% of the total in commercial fishing compared to the mean of about 1.3% among other species. Human fishing operations left 60% of the total; other species left a mean of 98.7% (Data from the 1980s, based on Livingston, 1993).

consonant with the question we can confine our-selves to unharvested production that is within the limits to natural variation in what other species leave for the rest of the ecosystem. Consumption and nonconsumption are dealt with consistently (Hobbs and Fowler 2008).

A third way to set priorities is to follow the lead of the medical application of systemic management. If a person's temperature and body weight are both abnormally high, priority for initial action will probably be given to the fever, even if body weight is equally abnormal. This decision would be based on knowledge of the known risks of imminent death associated with fevers. The problem of being overweight would not be left unattended, but the choice of which problem to address first would be made based on known risks. Alternatively, abnor-mal cholesterol levels and body weight might be treated on the basis of which one shows the great-est abnormality. That is, risk in some cases may be proportional to the degree that the situation is pathological. Similarly, reducing the amount of energy we are appropriating within ecosystems and the biosphere (Fowler 2008), would be given higher priority than becoming strict vegetarians to change our trophic level (Fig. 2.3), because humans are near the center of all frequency distributions for trophic level. Becoming vegetarians would con-tribute to solving the problem of energy appropri-ation but energy appropriation requires dealing with overpopulation to solve the majority of the problem and would be given priority both because of the magnitude of the problem and the degree to which it would solve other related problems.

A final means of setting priority may be found in information linking human abnormality to abnor-mality in the risk of extinction—not just human extinction, but extinction in general. This is an area where simulation modeling can conceivably play a very important role.

5.2.2 Species-level patterns (frequency distributions) are inadequately developed

When applying frequency distributions to human individuals, we distinguish the normal range of natural variability from the abnormal (aberrant, anomalous, pathological) range. Body temperature,

blood pressure, or body weight (Calle *et al.*, 1999) sufficiently outside the normal ranges of natural variation present abnormal risk. However, their occurrence cannot be characterized as unnat-ural, nor are the risks of being outside the normal ranges unnatural nor unreal. Frequency distri-butions for individual-level metrics (individual-level patterns) are emergent natural phenomena, just as are species-level patterns (or species fre-quency distributions). What we do not have for species frequency distributions, that we do have for individual frequency distributions, is the in-depth knowledge of their regularity, how they vary with other characteristics (as weight varies with age and gender), and what the magnitude, predictability, and nature of some of the risks are. Death is known to be associated with fever and the increased risk with increased fever is an accepted phenomenon. The precautionary principle dictates that, when we do not know, we act conservatively until we find out.[41]

Future research will improve our understanding of what is normal and what is abnormal at the spe-cies level. The risks of extinction that are associated with varying degrees of abnormality will become better understood through research involving both empirical information and modeling. The entire range of variability is natural and humans are part of natural variability. We are subject to the laws of nature that present risks that define normal ranges of natural variation. These are also natural. Being within the normal ranges of natural variability accounts for the complexity of factors that present the risks.

Dealing with complexity, however, is not a sim-ple issue, even with insight that shows how to account for it. If we look at systemic management as benchmarking,[42] it is *not* a matter of doing what any other individual species does, nor does it involve seeing an individual species (especially ourselves) as optimal in their sustainability. In nature's trial-and-error approach to finding sustainability, many trials are failures; over 99.9% of all species have ultimately failed. Optimality only hazily emerges in measures such as statistical central tendency or maximized biodiversity (Fowler 2008) for species-level patterns appropriate to humans and prevail-ing conditions.

Even collectively, groups of species as currently observed are not a perfect basis for defining ultimate sustainability for humans. Abnormal human influence is inherent to these distributions as we see them today. They reflect what is sustainable in the face of human influence. In view of the current extinction crisis, such sustainablity is quite tenuous. Our contribution to the environment experienced by these species has not been within the normal ranges of natural variability for some time. This is clearly shown in Figures 5.1–5.3, for example. If current population sizes of other mammals of our body size were not responding to the atypical levels (and effects) of the human population (and all other human abnormalities), they would serve much better as frames of reference, as approximated with the hypothetical distribution for such conditions in the top and bottom panels of Figure 5.1. Part of being adaptable involves changing the sustainability targets for our species as systems respond to changes we can muster (assuming we can make such changes in time to avoid human extinction as a systemic reaction to past and present abnormal human influence).

Many existing species frequency distributions do not adequately account for space. The consumption rates for walleye pollock in Figures 2.6 and 5.5 are from major portions of its geographic range—pollock populations. They do not necessarily represent consumption rates for the species as a whole. Estimates of portions consumed from the entire species would be necessary to produce a more useful species-level pattern if we are asking questions about our harvest from the species as a whole. Some species that consume walleye pollock may take very few (or very little biomass) because of very limited overlap in geographic ranges. This would be one of the many factors contributing to what we see (an e_i in Fig. 1.4 contributing to patterns such as that of Fig. 2.6). Alternatively, consumption by one consuming species may appear abnormal if measured only in an area of overlap and not applied to the resource species as a whole. For example, such problems could be more extreme for the pattern in consumption rates of fish species of the Northwest Atlantic (Appendix Fig. 2.1.5) than for walleye pollock if they represent a smaller fraction of the overall geographic range of the prey species.

Another reason we want to be cautious in the use of existing species frequency distributions is the nearly total lack of temporal variability (at all time scales) of these distributions. Assessing sets of species, as well as individual species (especially humans) depends on such information. Consistency in the shapes and positions of such distributions for other systems, across space and time, will help evaluate the utility and correlative analysis of any individual distribution.

In fisheries management, for example, the walleye pollock fishery would be managed by constraining our take to less than 15% of current harvest levels if we applied existing information directly (Fowler 2008). While beginning by making large reductions in the take is undoubtedly advisable, a full 85% reduction may either be too much or too little.[43] We would be much more confident in suggesting such radical management if we had 30 other sets of fishery data like that for walleye pollock covering the entire geographic range of the resource species, all indicating very similar optimal take rates. We would be even more confident in such a recommendation if 10 separate sets of data, spread over decades of time, for walleye pollock in its complete geographic range, showed the same distribution (even if data for the consumer species individually showed them to reposition themselves within the histogram representing the full set of walleye pollock consuming species). Similarly, data for predators feeding on species otherwise similar to walleye pollock would be informative.[44] Consistency in cross-system comparisons and consistency in time would improve the foundation for decision making. How do such distributions vary over evolutionary time? In all cases, covariants such as body size, life history strategy, trophic level, and geographic range are helpful in directly accounting for such factors as was done for body size in looking at population density (Fig. 2.31, Fowler 2005).

The need for data and research is clear. This is particularly obvious for the multidimensional relationships mentioned in Chapter 2 and above. Interspecific multidimensional patterns are increasingly recognized (e.g., Brown 1995, Charnov 1993, Gaston and Blackburn 2000, Peters 1983, Rosenzweig 1995). Metabolic rates in relation to body size are clearly quite linear in log-log space. Clarity is developing in regard to the density/body size relationship (Damuth 1987, 2007, Schmid *et al.* 2001) and population variation/body size relationship (Sinclair 1996). Not all relationships can be expected to be so clear. Any relationship between total (global) population size and rate of increase has yet to emerge through direct observation, even though we expect one based on body size/density and body size/rate of increase relationships (see Appendix Fig. 2.1.22). Local density and range size appear to be related (Hanski and Gyllenberg 1997), but what about geographic range and rate of resource consumption? What about all the many multidimensional interrelationships yet to be examined (Table 2.1)?

The need to be adaptable is crucial to any approach to management, emphasized by the prevalence of adaptable species (the abundant smaller bodied species can evolve more rapidly, on the whole, than their larger bodied counterparts). Behavioral adaptations are within our options if we can use the information content of species frequency distributions. However, being adaptable includes being wary of the limitations of even these rich information sources. The changes we have set in motion, at all levels of organization and time scales, are changes we want to be prepared to accommodate. These are in addition to the kinds of changes nature will present to us that are relatively independent of our influence (e.g., periods of glaciation, shifts in the magnetic pole, cycles in solar radiation, or continental drift).

5.2.3 Implementation requires addressing component questions

Using empirical information in systemic management establishes goals and objectives. However, knowing an ultimate objective does not specify the means of achieving it. While this is a limiting quality of systemic management, it is certainly not a problem devoid of solution—using systemic management. The more detailed aspects of change involve the need to address more specific questions. Many such issues can be addressed with measures and information that science can produce. However, complexity is not always a matter

of objective scientific insight. There are also issues requiring systemic management that effect such things as religious beliefs, designing political systems, formulating educational programs, or altering existing law in order to achieve sustainability—most of which do not lend themselves to precise scientific measure. They are subjective but systemic, nevertheless.

Thus, evidence of a fever does not contain advice as to how (or whether) to achieve a reduction in body temperature, any more than evidence that the human population is too large provides guidance on how (or whether) to reduce it. Refined components of each issue (refined questions, Appendix 5.2), however, lend themselves to the application of systemic information in setting goals to be achieved (Management Tenet 9). For example, we can systemically rule out ways that will not work (e.g., birth rates of zero would be below the normal range of natural variation for birth rates, if we were to decide to take action to reduce the human population through controlling birth). Systemic management would prohibit such extremes. In all cases, we have the benefit of the more significant contributions of systemic management: identification of goals to be achieved, situations to avoid, measures of the problems before us, and addressing subsidiary questions of implementation.

Thus, identifying problems, having a basis for assessment of status and needs, and setting goals do not solve the problems. These are important steps, but knowing what to do is different from knowing how to do it. Both are different from actually doing it (praxis). Navigation requires three things: current location, destination, and course to follow. With specific species-level patterns we have information about the first two. The latter requires refined component questions, and relevant patterns—including those for individuals. Experience and role models of what to do (praxis) are important at all levels.

How do we reduce the population of our species? If agencies that regulate our interaction with resource species (usually known as resource management agencies) could restrict our take of resources to normal levels in comparison to other species, the human population would decline because of reduced food supplies. Such a program could not be implemented without unprecedented coercion (or certainly unprecedented cooperation) on a global scale. No institutional or social structure exists wherein regulating our use of resources can be carried out to reduce the human population so that demand does not exceed supply (the issue is usually avoided as a matter of policy). How can management address the religious, economic, cultural, psychological, and social complexity of such change? How can community-based management (Western and Wright 1994), elected officials, businesses, family planning, and other religious, social, and economic entities be enlisted in reducing population? A species frequency distribution for population level can be used to set goals, measure problems, and assess our species and any progress toward achieving goals, but does not help devise implementation strategies. These are left to other aspects of systemic management, particularly refined forms of related questions, and include different levels of biological organization.

Doing so, it is all too easy to reach the conclusion that we cannot purposefully reduce our population through human action. It may be inadvisable; few if any other species purposefully regulate their populations. The alternative is that of accepting the forces of nature that regulate the populations of other species. These include disease, starvation, predation, and intraspecific strife. In nature this is the pattern. In nature, such forces involve natural selection. Carried out through human action, it is clearly impossible to avoid abnormal selectivity in such management. Joseph Meeker (1997) said: "The origins of environmental crises lie deep in human cultural traditions at levels of human mentality that have remained unchanged for several thousand years. … Given such depth, it is possible that "solutions" are more than can be hoped for. Humanity may have to settle for the distinction of being the first species ever to understand the causes of its own extinction. That would be no small accomplishment". We may experience a population reduction caused by a massive pandemic, or combinations of other major forces (that we would judge to be a catastrophe), rather than extinction. However, the systemic forces of nature ultimately prevail. The species-level patterns we have to work with indicate that a huge change is both likely and necessary for the restoration of sustainability.

5.2.4 Acquiring information is logistically difficult

Existing information on species-level patterns is the result of decades of expensive, difficult research. A species frequency distribution such as that for the consumption rates of walleye pollock (Figs 2.6 and 4.1) requires millions of dollars to produce. The sampling of stomach contents or feces from predatory species to study their diets involves hours of labor by many technicians, and years of study. Because of logistic difficulties, more precise estimations of population numbers do not exist for the species in Figure 5.1 and Appendix Figure 2.1.22 and are completely missing for many species. Collecting such data is not easy and all data are subject to measurement error. Luckily, libraries and databases maintained in numerous laboratories contain a great deal of useful information that has yet to be synthesized into species frequency distributions. Most, however, were collected for other reasons and suffer the inadequacies of many sets of data such as those shown in this book. Despite the difficulties, the potential benefits for management justify the importance of effort to acquire the best information we can bring to the tasks before us. The production of such data sets constitutes the science best suited for management (Belgrano and Fowler 2008).

5.2.5 Statistical and biodiversity measures are reference points—not magical numbers

One of the lures of maximum sustainable yield approaches to managing our fish, wildlife, and forestry resources has been its simplicity, especially a single number that emerged as a solution to an equation. Statistical measures of central tendencies, and biodiversity maxima (Fowler 2008) among other species in species frequency distributions can be misused in the same way. They are often likely to be close to optimal circumstances. They are certainly much better than anything outside the limits shown by species-level patterns free of abnormal human influence. However, we should avoid seeing these reference points as simple end points for our management for a number of reasons.

First is the bias in our depiction of most existing species-level patterns (e.g., frequency distributions) because of atypical human influence, mentioned numerous times earlier in this book.

Second is a lack of clarity stemming from the preliminary nature of existing samples. Temporal variability, cross-system variability (among ecosystems, taxonomic categories, or any other species sets), and variability in multidimensional space are important to work out in greater detail, over both time and space.

Third is the difference in measures of central tendency themselves for attributes that show varying kinds of statistical properties. Means and modes are very rarely the same in variations that are often, but not always, log-normally distributed and often dependent on how the variable being measured is defined (e.g., tons of biomass harvested vs. portion of standing stock of a prey resource consumed). Central tendencies or statistical limits may differ from positions for humans that would maximize information (Fig. 5.3, Fowler 2008).

Fourth is the difference in degree to which various contributions to the formation of species frequency distributions come from extinction, speciation, evolution, and nonevolutionary factors such as ecological mechanics. For example, the concentration of species in small geographic ranges, producing large "refuges" from the direct effects of each species (Fowler and Hobbs 2002, 2003), could result from evolution of habitat specialization, or it could come from the extinction of all species with large ranges. Producing information to distinguish among each of the contributions will be an endless task. In the meantime, knowledge that factors such as connectivity lead to instability in ecosystems reminds us that being an extreme outlier is not advisable.

Fifth is a matter of "time-in-place"—how long a species has existed. A species at the edge of a species frequency distribution that has been there for two million years may be a better example of sustainability than a species in the middle (e.g., at the mean) for only a thousand years. Turnover will be impossible to account for in the foreseeable future. The weight we would assign to a species according to the time it has shown a particular characteristic would be expected to vary across the spectrum of values measured in producing species frequency distributions. Ultimately, sustainability does not seem to be an option outside the normal ranges of natural variation. It is impossible to find evidence of sustainability, even where positive weightings might exist, in regions of species frequency distributions not occupied by species.

Sixth involves the effects of the physical environment. Part of the variability we expect to see among species frequency distributions for a particular species characteristic is related to the mechanical and evolutionary effects of the physical environment through selection at all levels. This means that we can expect to see differences in species frequency distributions depending on such factors as location, habitat, temporal variation, and chemical composition of their physical environment (e.g., soil type and water quality). These factors must be considered as covariates in the use of species frequency distributions, similar to establishing harvest rates of fish by using mean water temperature, insofar as temperature might influence the position of central tendencies.

Finally, measures such as information indices (Fig. 5.3, Fowler 2008) may better reflect sustainability than do simple measures of central tendencies. Being within the normal range of natural variation is a clear objective when we are so far removed, and maximizing the biodiversity of systems may be a way forward. For many situations it will require less change than would matching ourselves to means or other central tendencies of species frequency distributions (Fowler 2008). Again, however, fixed points are not what we are after as there is the matter of normal variance in all cases.

Most, or all, of the concerns related to the factors listed above can be dealt with empirically, through correlation analysis and modeling in multidimensional space, based on empirical interrelationships (i.e., not models restricted to ecological mechanics or simulation models). However, the databases to carry out such research are limited and very difficult to produce. The identity of these data bases depends on asking both the correct management question and the corresponding (consonant, isomorphic) scientific question (Appendix 5.2).

5.3 Why extinction should be a management issue

In the history of the Earth, billions of species—over 99.9% of all those that have ever existed—have gone extinct. Statistically, no species' chances for avoiding extinction are high and the human species is no exception. Another ice age may occur, or asteroids may strike the Earth, drastically reducing or eliminating the human (and other species') population within days or decades. Chances are high that we eventually will go the course of most other species. Ultimately, when our sun undergoes the changes astronomers project, life on Earth will likely be impossible. Aside from such risks, which are beyond our control, the abnormal, extreme position of humans in comparison to other species in so many species-level patterns (Fowler 2008, Fowler and Hobbs 2003, and as will be seen in Chapter 6) justifies action—management by humans (and of humans, Management Tenet 2) to prevent abnormal risks of human extinction owing to anthropogenic causes. Simultaneously, of course, we would be striving to minimize any abnormality in the risk of extinction among other species owing to our influence.

In Chapter 3, it was briefly argued that extinction is a more significant force than evolution or the nonevolutionary factors of ecological mechanics in contributing to the formation of patterns among species. This will be the subject of intense scientific debate for some time. The validity of such an argument may never be proven. However, the burden of proving (or disproving) it shifts in systemic management. Those who maintain that extinction is not a risk worth being concerned about now have to prove their case. We are concerned about it for other species, why not for ourselves? Systemic management levels the playing field. If it is not important, the matter of proving so is shouldered by those who support that position in their decision making while scientists debate the issue from both sides (Mangel *et al.* 1996). In the mean time, its *actual* importance is taken into account (Fig. 1.4) in systemic management and the debate becomes a moot point. Nevertheless, some relevant issues are worth considering.

The argument that extinction is an important factor to consider in management may be summarized as follows. The processes of extinction are hierarchically inclusive of natural selection at the individual level such that what is good for the part may be lethal for the whole (Table 3.1, Appendix

3.2). Both forms of selection are inclusive of ecological mechanics.[45] Human concerns should not be limited to the risks of ecologically mechanical reactions of ecosystems to our influence (e.g., diseases and starvation). It is likely that extinction supersedes these as a controlling or constraining factor, even though disease and starvation can be involved in the complex of factors behind extinction. It follows, therefore, that extinction is probably more influential in forming and maintaining species frequency distributions than are other forces. If this is the case (consistent with the behavior of models like those of Chapter 3) we would be advised to make extinction a priority management issue. Our concern would not be confined to the risk of extinction for other species, but would include all species (i.e., would not ignore humans, a requirement of Management Tenet 1).

Precautionary approaches are to be taken in the face of uncertainty. We are uncertain about the degree to which extinction is involved, but we know it is involved. The pattern in hierarchical constraint seen in complex systems (Ahl and Allen 1996, Campbell 1974, O'Neill *et al.* 1986, Wilber 1995) is basis for precaution. Extinction (as with all forms of selective systems failure) is involved in emergence at each stage in the history of hierarchical development (Morowitz 2002). Extinction is incorporated in the application of intransitive management guided by species-level patterns. Regardless of its conceptual importance, it is considered in proportion to its effects and actual importance in guiding information based on empirical information—automatically, inherently, and systemically. Systemic management is precautionary in this regard, and not just in regard to the risk of extinction. No precaution based on guesswork, opinion, or political stance is justified because all risks are accounted for collectively, especially in avoiding the abnormal in all cases wherein we find ourselves to be atypical.

5.4 A protocol for systemic management

The guidelines below are intended to provide a very brief overview of the steps in implementing systemic management. They are based on the management tenets described earlier, primarily in regard to setting goals for species-level issues, including our species' interactions with other levels of biological organization. Implementation forces consideration of issues at other levels—issues that can also be approached systemically.

One overarching objective of systemic management is to alleviate our species, individual humans, and all other biotic systems of the effects of abnormal human species-level attributes (Christensen *et al.* 1996, Fowler and Hobbs 2002, Mangel *et al.* 1996). Optimally, this involves finding the ways to participate in larger systems that minimize any abnormality of risks. This avoids the inherent risks of extremes, including extinction as one of the risks all species face (Chapter 3). Observed limits to natural variation among species, seen through research revealing species-level patterns, serve as preliminary standards of reference (Fowler 2008). They embody consideration of complexity, including limits, through their Bayesian-like integration of the complete suite of factors species encounter in their participation within ecosystems and other larger systems (Plate 5.2).

In very abbreviated and oversimplified form, the protocol for management to achieve sustainability (with focus on involving the human species) is:

1. Define a management question (and define others through subdividing, resolving, refining, and expanding).
2. Identify relevant levels of biological organization (species in most of the examples of this book) and associated characteristic(s) to be measured.
3. Find the limits to the normal range of natural variation through research on the identified characteristic(s).
4. Take action to avoid human abnormality wherever abnormality is discovered (change our species when it is a species-level question).

There is an extremely important shift in the involvement of stakeholders to be noted at this point. This was depicted in Figure 1.1; stakeholders no longer convert information, biases, political positions, emotions, partially insightful understanding, economic interests, or specialized agenda to an objective. That role is terminated. Instead they are involved in asking questions (both manage-

ment and research questions). This is perhaps the most significant difference between systemic management and conventional management, especially in the progress it makes in achieving objectivity. This objectivity is achieved at all levels: consonant reality-based management applies to its components—ecosystem-based, biosphere-based, or single species-based management. The key element involved in the process outlined above is that of asking questions: the management question and the science question must be consonant (isomorphic, with identical units, logical types and circumstances). A matter of consonance important to this book is the matter of species-level consonance—comparing species wherein we compare ours with others. When we ask management questions about our species, information about other species is consonant information insofar as it involves species-to-species comparisons. This is fleshed out in Appendix 5.2.

The process outlined above is repeated for every management question that can be imagined in accounting for complexity. Implementation requires addressing questions at all levels; stakeholders are directly involved in implementation (praxis). Below is an example of how this protocol can be applied to a fisheries management issue.

5.4.1 Defining the question

Questions that do not have to do with the human are rejected as management questions, at the outset. They do not involve action we can take and present situations impossible to control (Management Tenet 2); they are not directly applicable. Thus, a question such as "What is the sustainable level of a population of Steller sea lions?" is not directly applicable as we cannot control sea lion populations, although sea lions that show abnormality prompts asking questions about what we might be doing to influence them or their systems. This leads to directly applicable questions. Such questions include "What is a sustainable harvest of Steller sea lions?" or "What is a sustainable harvest of walleye pollock—one of the sea lion's prey species?" When dealing with an endangered species, we would ask as many relevant questions as we can. For example we would ask:

• What is a sustainable harvest of biomass (by humans) from the endangered species' ecosystem (e.g., the Steller sea lion as an endangered species in the eastern Bering Sea and North Pacific)?

• What is a sustainable harvest of biomass and numbers (by humans) from all prey species for the endangered species as a group (two questions, involving a species group as shown in the second panel of Fig. 4.1)?

• What is a sustainable harvest of biomass and numbers (by humans) from each individual prey species for the endangered species (top panel of Fig. 4.1)?

• What is the optimal allocation of catches (by humans) over time, space, and the species that serves as resources for the endangered species (three questions combined)?

• What is the sustainable rate of production (by humans) for toxic substances that find their way into ecosystems occupied at any time by the endangered species?

• What is the sustainable rate of production of CO_2 (by humans) that affects ecosystems occupied at any time by the endangered species?

The list of questions is endless and involves any concern held by any stakeholder.

In defining a management question, we begin to account for correlative information and complexity. Consider the question: "What is the most sustainable harvest of walleye pollock?" We can refine this question (see Appendix 5.2) to "What is the most sustainable consumption rate of walleye pollock biomass, in the eastern Bering Sea, during the summer, by warm-blooded species with a body size similar to that of humans, when the walleye pollock biomass is estimated to be seven million metric tons?" The latter version initiates consideration of correlative information. Such information includes species-level characteristics, both for humans and the other predators, and accounts for environmental circumstances. More can be added as further research elucidates more such correlative relationships. For this example such relationships could include sun spot activity if consumption rates of pollock by nonhuman species are found to show a correlation with sun spot activity. The refinement process involves reductionism; it makes

use of reductionism as one of the strengths of our thinking and its expression in science.

Component questions must also be asked—beginning the process of guiding the implementation of a goal involving total catch. What is the appropriate allocation of catch across age or size (Etnier and Fowler 2005)? When and where should walleye pollock be caught (two questions)? How should we allocate catch across sex? What is the optimal catch when measured in numbers? What is the most sustainable catch of walleye pollock at the age and size we wish to take (two questions)? These are questions involving objectives. Who enforces the resulting regulations? What fishing techniques, technology, and equipment are to be used? These latter two questions involve implementation. All add to the consideration of complexity through the involvement of component systems and respective management for consistent consideration of hierarchical organization (Management Tenet 4).

We must also expand the question (Appendix 5.2). The taking of walleye pollock adds to the harvest within the ecosystem, it adds to the take from various groups of species, it adds to our harvest from the biosphere, and it adds to the number of species that we are harvesting. Do these still fit within the normal range of natural variation when they include our predation on walleye pollock? In other words, other questions deal with whether or not it is appropriate to be harvesting walleye pollock at all. Thus, the meta-level questions are not ignored and are added to systemic management in further consideration of hierarchy though consideration of inclusive systems.

5.4.2 Identifying characteristic(s) to be measured

Ensuring the measurements are *directly* related to a specific management question is fundamentally important. Mere relevance does not suffice; the pattern produced in research must be *consonant* with the management question (Fowler 2008, Hobbs and Fowler 2008).[46] The concept of maximum sustainable yield became a problem in conventional management in failing to ensure such consonance (Fowler and Smith 2004). One set of information typically used in conventional management is the rate of population increase as a function of population size/density (directly leading to estimates of what has erroneously been called Maximum Sustainable Yield—MSY). This pattern is consonant with a different management question. The consonant management question is: "At what rate should the human population be increasing if it is at half of its normal levels?" The question in systemic management that people are trying to address with MSY is: "What is the rate at which we humans can most sustainably harvest a resource species?" The latter *is addressed directly* with information about rates at which resource species are harvested (consumed) by other species. Part of the consonance achieved with such information involves the species-level aspect of the management question, science question, and the pattern revealed by science. This goes on to include the action taken.

Thus, the management question defines the measurements to be made. In systemic management, the asking of management questions is one of two roles of all interested parties (bottom row of Fig. 1.1). The initial question (in the example above) regarding the harvest of walleye pollock leads to measuring consumption rates among all species of consumers that feed on this species—thus defining the best science for this management issue. Various units of measure are involved; for example, harvest rates should be measured in both biomass and numbers per unit time for the relevant area. Furthermore, measurements should be expressed in identical mathematical transformations. If comparisons (revealing abnormality) are based on log transformation, these need to be converted back to raw units for management. This may involve tons or numbers per unit time, Figure 1.7. Another aspect of management involves portion of the standing stock taken in harvests as the unit of management; log transformed values used for comparison must be converted back to the measures defined by the management question (e.g., fraction of the population of walleye pollock harvested, Fig. 2.6). In each case, if comparable measures of human consumption indicate that our species is outside the normal range of natural variation, we have a problem, and management action is needed.

If we start with a management question regarding the sustainable harvest of walleye pollock

as the resource species, a more refined question requires measures of biomass consumption (distinct from numbers of individuals, which would be addressed as a separate question). Further refinement involves such things as biomass consumption in a more specific geographic region, during a specific season, under specified environmental circumstances, at a particular level of the resource species population, by specified kinds of species, but still involving walleye pollock as the resource species. We get such information by measures that involve consumption rates among mammalian predators of human body size, collected during the summer, in the eastern Bering Sea, during conditions of high sun spot activity, and when walleye pollock occur at a biomass of seven million metric tons, where predation is measured in units of biomass consumed per unit time. If we want to harvest adult walleye pollock we would use measures of the consumption of adults by walleye pollock predators. This example illustrates the impossibility of accounting for everything directly in formulating management questions because we cannot list all of the factors that may be involved. However, it also illustrates the value of doing what we can and the progress represented in this process made by systemic management in comparison to conventional approaches.

5.4.3 Finding the normal range of natural variation

Field research and analysis of historical data provide information to define limits to variation consonant with the management question. The resulting information is the scientific information that best meets the need of management (Belgrano and Fowler 2008). The research that produces it constitutes the science best suited for management, thus addressing a long standing need in management (NRC 2004) in view of the limitations of science (Appendix 5.1). Logistical constraints prevent being explicit about many (most) issues involved. Where information is lacking completely, correlative information often offers an alternative to help inform by way of extrapolation and interpolation based on information from other systems, locations, and conditions. Comparisons across morphological, temporal, and spatial scales inform us about correlative relationships to directly account for such things as body size (see Peters 1983), season, and latitude. Thus, in addressing a question regarding sustainable catches in fisheries, body size of both the predator (humans) and prey can be taken into account directly in correlative sub-patterns involving consumption—consumption being the most important element of consonance.

To the extent that useful data are already available, the relevant frequency distribution can be used to define limits to natural variation. Further field research (with research questions defined by scientists and other stakeholders; bottom row of Fig. 1.1) adds to the collection of needed data. Any time existing data result in the conclusion that humans are outside the normal range of natural variation, management action is indicated (as in the case of harvests from walleye pollock, finfish, and the eastern Bering Sea in regard to the questions we are addressing, e.g., Fig. 4.1). If we are inside the range of variation exhibited by other species, we may be in a position of sustainability. However, we cannot know we are being sustainable under such circumstances because we cannot prove that we are. We cannot know exactly where, within the normal range of natural variation we should be to maximize sustainability, but can make gains in using correlative information in refining questions (see Appendix 1.3 and Fowler and Perez 1999, for guidance on producing histograms representative of species-level patterns as frequency distributions useful to management).

5.4.4 Taking management action

Management is most important in the action taken (praxis); prior to that everything involves preparation devoid of effect without action (i.e., merely words, data, discussion, research, plans, concepts). Management itself is action that is taken to change or confine human behavior, influence or characteristics to avoid abnormality. This is where the social, legal, institutional, ethical, religious, economic, racial, and behavioral aspects of human endeavor at all levels—from individual to international—are involved in achieving specified objectives. Part of the reason systemic management was given

the name involves the systemic quality of change required of humans. Note that human endeavor that is involved in specifying the objectives is restricted to obtaining empirical information and using experience from past trial-and-error processes; the setting of goals is taken out of the hands of stakeholders (in the top row of Fig. 1.1) and replaced with empirical information. Establishing goals, as an activity by stakeholders, is replaced with the role of asking questions; defining clear, consonant management and research questions. Even scientists are limited to gathering empirical data, displaying, and then interpreting the limits to natural variation rather than giving advice based on their necessarily human concepts, thinking, models, or consensus. In systemic management, human endeavor is less a matter of guiding and much more a matter of management itself—achieving the objectives specified by observed patterns.[47] Guidance is provided by what we observe and have experienced. The abstractness of words, ideas, models, concepts, and nonconsonant information make them important only insofar as they help lead to more management questions, result in correlative information (seeing correlative subpatterns), produce representations of relevant natural patterns, and undertake subsequent management action.

The simplifications made in considering the example of fishing for walleye pollock, dealing with endangered sea lions, and the eastern Bering Sea ecosystem above cannot be overlooked. Ideally, systemic management requires addressing all management questions, refined and expanded to the extent possible. The same protocol can be applied in the other realms and dimensions. In the Bering Sea, there may be concern about global warming; this gives rise to management questions about CO_2 production. Other concerns may relate to pollution which give rise to questions about the sustainability of production of the tens of thousands of manufactured chemicals. Interested parties might question the focus of harvests on particular species. This gives rise to questions regarding the sustainability of the numbers of species harvested, and the allocation of harvests across those species (selectivity among them). Some may pursue the concept of marine protected areas which leads to conducting research on the portion of the ecosystem occupied by consuming species (Fig. 2.15). On a more global scale, questions concerning the sustainability of our population size, energy consumption, water consumption, and geographic range size come to mind as we consider the state of the planet in general. All such issues can be considered separately in local regions, ecosystems and the biosphere. This list is only the beginning of another list that also proves impossible to complete. Managing systemically proceeds on all fronts for which we have the capacity to ask management questions and obtain informative, integrative, guiding information.

An analog with individual health would be to manage by maintaining pulse, respiration, body temperature, and metabolic rate while attempting also to, for example, bring body weight into the normal range of variability. We would attempt the same for our species. Each level involves refining and expanding management questions, identifying measures, determining the normal range or variation, and taking corrective action where abnormality is found.

In all cases, the implementation of systemic management must include individuals in solving management problems for our species. However, solving the problem of overconsumption of energy or biomass is not dealt with systemically by requiring people to subsist on a stalk of celery per day. It is important that we maintain ingestion rates, mortality rates, and birth rates that are within the normal range of natural variation. We cannot place the solving of species-level problems on the shoulders of individual people through short-term solutions. Only through long-term normal sacrifices at the individual level can we accomplish our objectives transitively. This is where systemic management becomes a matter of addressing individual-level questions to define the process of implementation. On all fronts, we must remain open to the option that the best and most complete solution will be found through systemic effects of ecosystems and the biosphere—solutions that will temporarily be abnormal for us (and judged in our value systems as horrific), but completely normal for the more inclusive systems involved.

5.5 The eastern Bering Sea example

As is now clear, systemic management applies at all levels (individuals through to the biosphere); systemic management is reality-based management and all are hierarchical parts of reality. In regard to ecosystems, it must be based on answers to specific management questions about specific aspects of human interactions with ecosystems (as described above; see also Belgrano and Fowler 2008, Fowler 2003, 2008, and Fowler and Hobbs 2002, 2003). These questions form part of the step-by-step process of systemic management based on our roles within an ecosystem, particularly as a species (but including individuals so as to avoid being confined to species-level management). Clearly specified management questions then require consonant data for guidance in avoiding the abnormal—data from scientific studies to represent the best scientific information available. In order to guide harvest practices under different environmental conditions such studies need to include data collected under various climatic regimes. Similar studies for other marine systems, as well as terrestrial, freshwater, desert, and alpine systems, will be helpful for inter-ecosystem comparative studies (and the correlative patterns involved).

How much biomass can sustainably be removed from the eastern Bering Sea by humans? The answer to this question lies within information such as that shown in Figure 4.1. In its current form, such information is limited compared to what we would have if there had been no abnormal human influence historically. It is limited to conditions insofar as they have responded to human influence. The ways to produce better information, and actions to take while waiting for its accumulation, are now clear. Scientists can now direct their attention to research that produces data consonant with the many questions facing managers—one set of data per question. These include questions such as: How many species can sustainably be harvested as resources? How do we allocate our harvests across space and alternative species? What portion of an ecosystem should be left as a reserve (or in multiple reserves)? These and other questions lead to needs for information (research, field studies, synthesis of existing information) about the limits to natural variation as illustrated in species-level patterns.

Within an ecosystem, our species interacts with other species; we influence each individual species. One of our more direct influences involves our harvest of resources. Measuring consumption by other species provides consonant information for influence that involves consumption. More consonance is achieved by more specificity—directed and intentional reductionism. How does predation vary according to prey density, trophic level, body size, age, and life history of the resource species? How does it vary according to body size, geographic range size, trophic level, and population density of the predators? The sustainable harvest of adult walleye pollock may be even less than indicated by Figure 4.1 when such factors are considered directly. More research is needed. The reductionism involved is a careful choice of the products of science—avoiding the misdirected reductionism of conventional management (top row of Fig. 1.1; Belgrano and Fowler 2008). Our conscious purpose becomes one of holistic design.

To address questions regarding sustainable harvest rates, scientists studying the eastern Bering Sea can emphasize studies exemplified by those of Livingston (1993) and Sobolevsky and Mathisen (1996). Similar studies by Overholtz *et al.* (1991) apply to the Northwest Atlantic. Studies such as that of Melin *et al.* (2008) apply to the Eastern Pacific and include information to directly account for correlative environmental conditions. The results of such studies achieve consonance in regard to management questions about harvest rates. Harvest rates involve consumption and measures of consumption do not have to be translated or converted to find management advice; there is consonance in units. Confined to simple measures of consumption, however, such information is somewhat superficial compared to what can be achieved, even with the limited data we have in hand today when we proceed to taking advantage of the correlative structure of patterns.

Figure 5.6, for example, shows the relationship between the harvest rates in commercial fisheries in relation to total mortality rate (M, from Mertz and Myers 1998) for 44 species of fish. In conventional management, comparisons between fishing mortality and total natural mortality are occasionally used to assess fisheries. As can be seen, many fisheries today are overharvested using this kind of conventional comparison (i.e., most of the solid points are above the 1:1 line). There is a significant lack of consonance here, however, in that commercial harvests represent consumption by one species; total mortality is total mortality—it includes the mortality caused by all consuming species. They are not the same thing; they are of different logical types. One is consumption per species (commercial harvests by humans), the other is the sum over all sources of mortality (which includes a variable number of predators each of which contributes to part of that mortality along with mortality caused by such things as diseases, accident, or cannibalism).

Correlative information regarding M (and thereby body size, intrinsic rate of increase, and other interrelated factors—Chapter 2) can result in refined consonant information found by plotting the mortality per consuming species (predation) against M. Doing so allows us to address management questions such as "What is the sustainable harvest of fish species with a natural mortality rate (M) of 0.4?" In Figure 5.6, this correlative information is based on data from consuming species (e.g., those represented in Fig. 2.6, and Appendix Fig. 2.1.5, also including data for the predators on three species of ungulates from Kunkel and Pletscher 1999). As can be seen, this results in a much different evaluation of fisheries; all fishing rates are above the line representing the relationship between predator-specific consumption rates and M. In other words, all fishing represented by these data are unsustainable by systemic standards. A biosphere-based approach to management (systemic management that would include all ecosystems and species) would result in harvests that would be much closer to the line representing nonhuman predators than is represented by the 1:1 line of Figure 5.6. Management might best be served by objectives based on a regression line of harvest (consumption per species) rates against M wherein the statistical fit involves consumption rates that maximize biodiversity for each prey species (Fowler 2008), rather than the simple rates of consumption for individual species as shown in Figure 5.6.

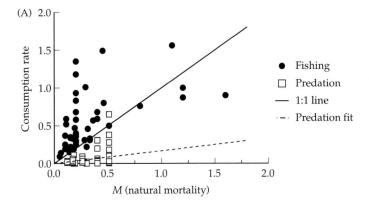

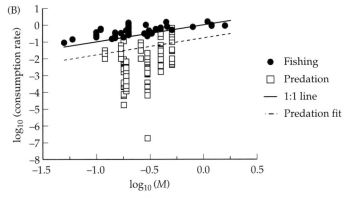

Figure 5.6 Two means of assessing commercial fisheries, one based on comparing mortality caused by fishing with total natural mortality (solid circles and solid line; Mertz and Myers 1998), and the other based on comparing fishing mortality with predator-specific mortality (solid circles and dashed line). The line for the latter involves a least squares fit of the data plotted in Figure 2.6 and Appendix Figure 2.1.5 (combined with data for ungulates from Kunkel and Pletscher 1999) to the corresponding total mortality rates for the respective species.

The pattern in Figure 5.6 raises the (scientific) question of whether or not there is an increase in the number of predators with increasing M. If there is such an increase, this would be a factor contributing to the nature of the relationship between mean predation rate and M, possibly making it a nonlinear relationship. The number of predators is obviously one of the factors that contributes to observed patterns, so is taken into account *a priori* in managerial use of the pattern. However, the number of predators would become a very important correlative variable, in any case, as refined research addresses such questions. We humans are adding ourselves to the count of predators in every case where we initiate harvests (such as done in commercial fishing). Increasing the number of predators involved must be part of what is taken into account with information regarding predator count as it accounts for variation in observed predation rates. Science and management involve reciprocity in question generation.

Still other questions relate to whether or not we can more directly take into account other processes, clearly relevant factors, and issues. For example: How do we account for the evolutionary impacts of our harvesting on each individual resource species? With consonant information, species-level patterns as natural phenomena provide a form of Bayesian integration (Chapters 3, 4, 5, and 6) that accounts for such issues. Thus, empirical information consonant with questions related to sustainable harvest rates already accounts for genetic effects (Fig. 1.4) as factors that contribute to patterns in consumption (harvest) rates. However, more directly accounting for genetic effects can involve both correlative analysis and the posing of new management questions. Thus, asking management questions about the sustainable size selectivity of fisheries harvests can lead to the use of consonant information on the size selectivity of marine mammals (e.g., Etner and Fowler 2005)—directly addressing evolutionary impact while simultaneously accounting for other aspects of complexity. Furthermore, selectivity may involve correlative patterns that allow us to account for other factors directly (e.g., body size, Etnier and Fowler 2005). Such correlative patterns could involve other forms of selectivity (e.g., patterns in selectivity by sex might involve correlations in which selectivity by size, location, or season are involved).

Systemic management also brings resolution to any differences in management advice stemming from different sciences by recognizing that the sciences are no more than (but fundamentally and critically important) elements of human perception of realities—realities that themselves are integrated (and given respective weights) in the combination of respective processes in their contribution to the various species-level patterns (frequency distributions, Fig. 1.4). These apply to any ecosystem, the eastern Bering Sea or the Serengeti.

With growing clarity, we understand correlative relationships within patterns as integral parts of emergent ecosystem characteristics. Ecosystems have real form and pattern to their function (e.g., Fig. 1.34). Correlative relationships provide the basis for taking various factors into account overtly or directly; when scientists/stakeholders are asked to consider a factor in management, it can be accomplished with empirical information. For example, we account for human body size and trophic level (attributes over which we have little short-term control) by finding our position within consonant species-level patterns as correlated with these features. Thus, we can establish a sustainable take of walleye pollock by comparing our harvest rates with those of other species of our body size and trophic level. We can account for climate directly if species-level patterns are correlated with seasonal temperature, climate regimes (as in the data provided by Melin *et al.* 2008), or annual precipitation.

Extinction is an imminent risk for endangered species. The Steller sea lion is designated as an endangered species in parts of its geographic range, including parts of the eastern Bering Sea. How would systemic management attempt to reduce the risk of extinction of the Steller sea lion? Because of the principle of complexity and interconnectedness, whatever we do affects other species in innumerable and unprovable ways (some positive and some negative when judged in human value systems). In systemic management, we would ask management questions about all of our direct and indirect interactions with the Steller sea lion and its environment. We would frame each interaction as a management question. We would ask about our harvest of sea lions, and any species the Steller sea lion uses as a forage species. We would ask about our harvests and impacts on all species in the ecosystems in which it occurs. We would ask about our production of chemicals that enter its environment, and CO_2 that may be changing the climate and pH of the waters inhabited by the sea lions and all other species with which sea lions interact. For each and every such question, we would follow the protocol presented above, beginning by correcting all the known problems, such as reducing our harvest of biomass from the eastern Bering Sea to levels within the limits to natural variation illustrated by data shown in Figure 4.1. For each and every component of our influence the full suite of consequences plays out in the complexity we account for automatically in systemic management. Only if we are within the normal range of natural variation of everything we can think of can we rest with any assurance that we have done everything within our power. Then, if the sea lions go extinct, it may still be a matter of past human influence or questions we do not know to ask, but we are powerless to change such things now. It is always possible that the sea lions are another species succumbing to the laws of nature and its dynamic process, regardless of any pathological anthropogenic influence. It may be, itself, experiencing the results of evolutionary suicide (Parvinen 2005, Rankin and López-Sepulcre 2005)—as an evolutionary dead-end (Potter 1990). However remote or important this possibility may be, it is not a reason for avoiding our responsibilities in avoiding the abnormal—one of which is the anthropogenic contribution to the current extinction crisis to be treated in more detail in Chapter 6 as it involves the complex or full suite of abnormalities characteristic of humans in today's world.

We cannot prove that what we are doing is causing reduced population levels of many, or even most, endangered species. To deal with complexity, systemic management amalgamates all such concerns to see the endangered species of the world as probable (not proven) consequences of abnormal biodiversity, resource consumption, chemical production, and energy monopoly (Fowler 2008). These, of course, are impossible to uncouple from the human population explosion (Fig. 5.1) expressed through the complete set of all related interactive processes and influences. These include our geographic range size, habitat appropriation, introduced species, monopolization of primary productivity, *and* those things we do not know to list, but for which we have the responsibility of correcting when the pathology comes to our attention.

Having established objectives for fishery harvests, systemic management then proceeds to questions of implementation where a great deal of past experience is brought to bear in addressing questions regarding gear type, technology, laws, regulations, enforcement, monitoring, economics, and other factors that involve individual fishers, regulatory bodies, and all relevant officials. Individuals participate in the management process by acting to help achieve established objectives and draw on personal experience to be involved with successful change (but avoid abnormal activities in the process). Experience shows how to successfully achieve established objectives and the exercise of the lessons learned through that experience is systemic management. If change toward established objectives is not being achieved, management is not being implemented. The details of implementation depend on past management experience; the objectives are established as described in this book.[48]

5.6 Summary and preview

This chapter has shown that systemic management meets all nine tenets of management. This is not surprising because it is defined by them. The requirements of each tenet are met while simultaneously and consistently adhering to all nine. It goes beyond these tenets to embrace other principles and laws such as the law of unintended consequences, and the principle of human limitations (being finite). It fully embodies reversal of the burden of proof.

Simultaneous consistency is a hallmark of systemic management (Hobbs and Fowler 2008). Through systemic management, human factors, including economic factors, are accounted for objectively in defining sustainable roles for humans in ecosystems and the biosphere to adhere to Management Tenet 1. Control is restricted to situations where it is a most likely option—control of the human to meet Management Tenet 2 by using guiding information (Management Tenets 5 and 9) that meets the requirements of all nine tenets. Empirical information accounts for complexity consistently when humans are held within the normal range of natural variation to adhere to all tenets. This rejects the concept of conventional management in which it is assumed that humans can adequately generate guidance in other ways (top row, Fig. 1.1). Systemic management meets Management Tenet 3 (accounting for complexity/reality) through the integration and consistency of factors achieved in nature used as the source of information. Complexity is addressed in a protocol that breaks down the process and requires that component questions be addressed, that other questions be addressed, that expanded questions be addressed, and that correlative information be used. Consistency is achieved, not only across the various levels of biological organization (Management Tenet 4) but also among the questions being addressed—all by confining humans to within the normal range of natural variation (Management Tenets 4 and 5). Risks (Management Tenet 6) are part of the complexity addressed in systemic management by using information (Management Tenet 7) on the normal range of natural variation (Management Tenet 5) as an accounting of complexity (Management Tenets 3 and 4). Science (Management Tenet 8) is fundamentally important to observing the normal ranges of natural variation relevant to each management question, monitoring progress in achieving success, and identifying problems. Through the data collected, goals and standards (Management Tenet 9) emerge in observed limits to natural variation (Management Tenet 5) to guide human action (Management Tenets 1 and 2).

This chapter shows that the failures of conventional management identified in Chapter 4 are largely overcome by systemic management. The finite of the human prevails, however; the finite of human nature is one of the elements of complexity dealt with in systemic management (Management Tenet 3). Because of this reality, we remain limited in our capacity to ask all the relevant management questions. This is one of the main challenges for the future: asking the right questions—taking our role in the bottom row of Figure 1.1 seriously. Also crucial is obtaining the consonant information (matching natural pattern and data with the management question and ending in consonant action). This is essential for finding guiding information regarding specific activities necessary to achieve goals—consonant implementation. What may be the main contribution of systemic management is the capacity to address the right questions whether they involve individuals, families, society, our species, or the interactions between these systems and other species, ecosystems, or the biosphere. For every question represented by measurable counterparts in nature, using empirical information for a complete accounting of complexity (Fig. 1.4) provides complete compliance with all nine management tenets. The more questions asked, guiding information found, and action taken, the more complexity has been taken into account. Solving one problem contributes to solving related problems, including problems unknown to us at this time.

This chapter has also shown that human limitations prevent exact precision for what is sustainable within the normal range of natural variation; the variance of variability requires that we do not try. We can never identify nor obtain information for all necessary correlative information. This need not be a particularly troublesome limitation because variation itself is one of the features of natural systems. Nevertheless, by using nature's examples of sustainability, with as much correlative information as we can obtain, we refine the objectives for achieving sustainability, not only for ourselves but for the systems upon which we depend (Management Tenets 1, 4, 5, and 9). The goals are embedded in the measures of pathology

we currently exhibit. The daunting challenges we face can now be measured objectively.

An academic treatment of problems such as we will see in Chapter 6 contributes to an objective view of our precarious situation and measures of needed change. Guiding information is to be found in directly related role models when the appropriate questions are asked; when full consonance is achieved between management question and empirical information. These realms represent the fertile ground for asking more questions, seeking the component questions, and proceeding with implementation based on ever more detailed component questions. However, if the expanded, higher-level questions are not asked, or if the guiding information is ignored, systemic management is not being conducted and cannot be effective. This has been one purpose of this book—to show how to pose questions above the individual- and species-levels of biological organization, in a way that can be addressed objectively. Full systemic management, however, cannot ignore individual-level applications; this book provides focus on the species-level to exemplify the process.

However, objectivity is not enough. Where do we find hope after seeing that we are in the deep trouble we will see in Chapter 6 (Fowler 2008)? Social, psychological, emotional, and religious issues are at stake as well as multiple belief systems. Economic, political, educational, cultural, and philosophical issues are involved. The complexity of human civilization, and the forces of our history, habits, and hormones cannot be ignored. In each and every such realm there are innumerable detailed concerns. Such complexity is systemic.

After adopting systemic management in principle, we find a natural expansion from the objective to the subjective—fully embracing sustainability in spirit. We carry within our genes the seeds of our own destruction (Table 3.1, even the capacity to obliterate life on Earth). We also carry the seeds of promise—issues, treated by, but beyond the scope of this book. These provide fertile ground for extensive development.

CHAPTER 6

Humans: a species beyond limits

We have today to learn to get back into accord with the wisdom of nature....

—Joseph Campbell, *The Power of Myth*

This chapter examines what is necessary to fulfil the requirements of Management Tenets 1, 2, and 5: avoiding abnormal human participation within ecosystems and the biosphere. This chapter also provides examples of management questions, guiding information, and goals or policy that are all consonant (match, are consistent and isomorphic). These examples show how consonance between science and management dissolves the barriers between them and makes conversion of information unnecessary. Some of the questions posed in previous chapters are answered; new questions are posed and answered. This chapter illustrates the staggering immensity of the problems we face and the goals to be sought in solving them.

The first section illustrates patterns that demonstrate the abnormality of humans and preliminary approximations of goals for related systemic management. As a species, we are extremely aberrant in our use of water, energy, and resources, in our production of CO_2, in the size of our geographic range, and in the size and density of our population. Other patterns add to this list, and this list will grow in future research to show more extremes of our anomalous nature. In systemic management, these empirical patterns can be used to guide action toward relieving various systems of abnormal human influence (Management Tenet 5). In essence, this would increase our species-level fitness and the sustainability of other species, ecosystems, and the biosphere. In later sections, the effective use of empirical information is described by way of example, with more detail regarding systemic management applied to problems at all levels of biological organization— individual, species, ecosystems, and the biosphere.

These examples demonstrate how systemic management accounts for complexity (Management Tenets 3 and 4), in part because ecosystems and other more inclusive systems (including their components, their processes and interrelationships, and their aggregate complexity) are all among the factors that both show their own limited variation and contribute to the limited variation we see in all systems. This chapter has the objective of implementing our understanding that the emergence of patterns involves all contributing factors; empirical patterns account for all contributing factors— reality, and all systems and dynamics that are part of reality. The focus is on shifting toward sustainability by regulating human interactions with other biotic systems. All such systems will respond to our actions.

Finally, more detail is provided for using systemic management to find sustainable levels of human influence in the eastern Bering Sea.

6.1 Limits to natural variation as limits to sustainability

In considering the needs of the human species, questions of sustainability immediately arise. What level of biomass can we sustainably remove from an ecosystem (e.g., for food or fiber; see Fig. 4.1)? How much should be left for other species? We need space but how much space can we sustainably occupy as a species. All species need space; how much space should be protected from direct human influence? How many species can be sustainably used to meet our needs; how much biomass can be sustainably harvested from any group of species? How should we allocate harvests of resources across alternative species? How much energy can we sustainably use and how much primary production should be left for other species and their ecosystems? How much CO_2 can we sustainably produce? This list will grow longer as we face the complexity of the reality with which we are dealing.

Questions such as those above include both ecosystem- and biosphere-level issues. As recognized historically, they also include our interactions with nonhuman species, individually and collectively. As for any species, meeting our needs requires other species, ecosystems, and a biosphere that can supply what is needed by all species—sustainably. What portion of the energy captured, and biomass produced, in ecosystems and the biosphere, should we leave for those systems and the species involved? We depend on ecosystem services. How do we ensure that every species is sustainably provided with the variety of ecosystem services upon which all species depend? We must adjust our needs to correspond to what other species, ecosystems, and the biosphere can support so that they retain sustainable qualities, services, and interrelationships. In doing so, we have the responsibility of addressing the disruption to other species, ecosystems, and the biosphere that have been attributed, or are attributable, to human influence (Appendix 6.1). For example, pollution is recognized as a global problem. This observations leads us to questions such as: "At what rate can estrogenic compounds be sustainably produced?"

It is important to find balance between benefits and risks both for meeting our needs and solving problems among other species, ecosystems, and the biosphere. Management Tenets 3, 4, and 7 require that policy and action be based on information that accounts for the complexity of reality to balance advantages with costs. When systems such as other species, ecosystems, and the biosphere exhibit abnormal characteristics, it is crucial that we identify any potential anthropogenic causes, direct or indirect—human abnormalities. Atypical human factors are open to being changed—action that will be to the mutual and ultimate benefit of all systems. Hence, we begin by noting that many of the problems other species experience, most known ecosystem-level problems (Appendix 4.2)

and biosphere-level problems (such as global warming, extinction, pollution, oceanic acidification) involve elements that are clearly of human origin (Appendices 6.1 and 6.2). Reversing the burden of proof leads directly from the observation of these problems to management questions regarding the sustainability of human activities—any that may be contributing, directly or indirectly. We are lucky, in this regard, because we stand a chance of solving problems measurable as human abnormality. Fixing/repairing other species, ecosystems, or the biosphere is something that would be highly unlikely (probably impossible) otherwise (Management Tenet 2).

The following sections offer examples of sets of information produced by the biological sciences that have practical application. They involve comparisons between humans and other species in regard to: resource use, CO_2 production, energy use, geographic range, and other issues—all of which are closely tied to the rapidly growing size of the human population, belief systems in conventional management, and resulting environmental degradation. This section illustrates the use of information on the limits to natural variation in addressing a variety of questions aimed at matching our needs to the limits of various systems to sustain us without eroding their capacity to both sustain us and to sustain other species. The picture that emerges is sobering. However, with understanding and an appreciation for the actual magnitude (Management Tenet 9, Chapter 1) of our problems comes hope for change that will solve problems—problems that include the current extinction crisis and any abnormal risk of our own extinction.

6.1.1 Use of biological resources

Humans are consuming biological resources at rates that are exceedingly aberrant in comparison to the rates at which other species consume the same resources (Fowler 2003, 2008, Fowler and Hobbs 2002, 2003). This section develops in some detail how patterns can be used to assess current consumption rates and identify goals for achieving sustainability. Commercial harvests of fish are used as one example because the methodology of systemic management has been most thoroughly developed for regulating our use of these resources (e.g., Etnier and Fowler 2005, Fowler 1999a, 2002, 2008, Fowler *et al.* 1999). Another example involves the sport harvest of ungulates. Evaluations have yet to be completed for consumption of other groups of species, for example, tree species used for lumber, species of mammals treated as game, and species of waterfowl taken for food and sport—not to mention agricultural species. It is important to emphasize that the impact of human resource use is magnified because, through global trade, we move harvested resource materials to locations well outside the natural geographic range of the resource species rather than recycling them *in situ*. Our impacts are further magnified by other factors related to resource consumption, such as the number of species consumed, impact on population size of resource species, and evolutionary impacts through genetic engineering, selective harvesting, and breeding. Each issue gives rise to a distinct management question. Scientific research provides systemic management with information about a pattern that matches (is consonant with) each question (Belgrano and Fowler 2008, Fowler 2003, Fowler and Smith 2004).

Food is primary among resources needed by consumer species, including humans. Management of harvests for human food supplies include those taken from individual species, taxonomic groups, communities, ecosystems, and the biosphere. Managers are responsible for achieving sustainable consumption rates regardless of the level of biological organization being addressed—rates that are consistent across the various levels of hierarchical organization. Managers regulating harvests from a single resource species, for example, must ask, "What levels of harvest are sustainable?" This question is insufficiently addressed if it is not also applied to ecosystems and the biosphere. The recognized failure[1] of quantitative population models, or models of productivity, in managing single species as sources of consumable products (such as food) has stimulated intense pressure to advance to ecosystems and the biosphere in broader approaches to management. To be more realistic, the entire suite of interactions must be taken into account in making management decisions. These

interactions include genetic effects (Conover and Munch 2002, Ehrlich 1989, Etnier and Fowler 2005, Law *et al.* 1993, Mangel *et al.* 1993–2005).

With systemic management, managers are able to account for complexity so as to include all the elements important to "ecosystem-based management". Currently, this complexity is not incorporated in the deliberations that managers or any other stakeholders undertake (the conceptual alchemy of the top row of Fig. 1.1). In systemic management, complexity is addressed through using information on the limits to natural variation (Fowler 2003, Fowler and Hobbs 2003, Management Tenet 5) in integrative patterns (Belgrano and Fowler 2008). Each individual species in species-level patterns is a provisional example of success in facing all factors involved in the spatial, temporal, and hierarchical complexities of reality with the interactions among all components (Fig. 1.4). By definition, these factors are the environment within which evolution[2] occurs. In the sense of adaptive management (Grumbine 1997, Mangel *et al.* 1996, Moote *et al.* 1994, Walters 1986, 1992), other species have passed the filter of risks and collectively reveal advisable forms of sustainability in their Monte Carlo (trial-and-error; see Chapter 3) processes. Evolutionary processes (including selective extinction and speciation) have contributed to the origin of empirically observed sustainable levels of harvest that can be taken from their resources.[3] What we observe are the results of all contributing processes (Fig. 1.4)—results that account for the full spectrum of ecosystem interactions we would consider important if we were capable of perfect knowledge.

6.1.1.1 Single species approaches
In accounting for complexity, management must include regulating our interactions with other species on a species-by-species basis—one part of the reality-based nature of systemic management. Our use of other species as resources serves to introduce the concept. Other interactions with species (e.g., genetic effects, population suppression, or redistribution in geographic space) cannot be neglected and should be treated in the same way with their own distinct management questions.

The first example involves the management question: "What is a sustainable harvest rate for fisheries in the take of walleye pollock (*Theragra chalcogramma*)?" This involves resource use as human predation on a species of fish. Figure 6.1 shows the pattern for predation rates on this species by a variety of consumer species in the 1980s. This figure is a modification of Figure 2.2 to compare the take of walleye pollock by U.S. commercial fisheries to predation rates by other species. The amount of walleye pollock biomass recently harvested by humans (top panel) is 39 times larger than the arithmetic mean of that by other species (bottom panel), over 200 times larger than the mode of the consumption rates by other species and over 1.5 times as much as the total consumed by the other species represented in this sample. One option for achieving sustainability would be to take the data behind

Figure 6.1 at face value and apply systemic management by reducing harvests of walleye pollock to 2% of recent levels (close to the arithmetic mean for the other species). There are other less extreme options (Fowler 2008), but none would have allowed for harvests larger than 20% of what has been taken on the basis of the policy-setting process of conventional management (top row Fig. 1.1).

Figure 6.2 shows such a reduction (from recent rates represented by the dashed line on the right to the solid line at the peak of the curve in the

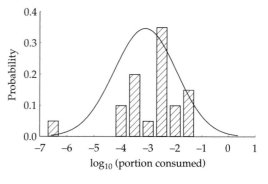

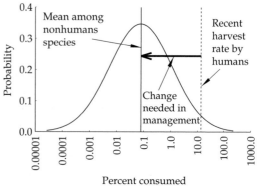

Figure 6.2 Illustration of the change that would be needed through systemic management of the harvest of walleye pollock (age three or older) in the eastern Bering Sea as a preliminarily maximization of sustainability. It uses the mean of the distribution shown in Figure 6.1 to establish a simplistic management goal (the mean is the peak of the curve in both panels even though such smooth curves rarely capture the complexity of such patterns; e.g., a symmetric curve may be entirely inappropriate). Management would be actions taken to reduce harvest rates (from levels represented by the dashed line) to correspond to the mean among the nonhuman species (represented by the solid line), as shown in the bottom panel—a change of about two orders of magnitude—as a first approximation of change needed in management with the objective of sustainability.

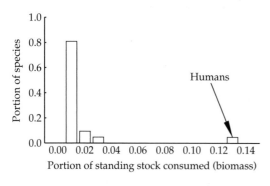

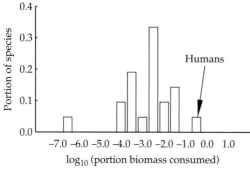

Figure 6.1 Frequency distribution of vertebrate species (N = 21), including humans, that consume walleye pollock (*Theragra chalcogramma*) in terms of the fraction of standing stock biomass of that species that were consumed in the eastern Bering Sea in the 1980s. Data are shown in linear scale in the top panel and in $\log_{10}$ scale in the bottom panel for walleye pollock of age three and older (Livingston 1993 and personal communication). The total take by humans represented here is about one million metric tons.

Plate 1.1 Water has a variety of physical properties and exemplifies one of the many substances that flow among the components of ecosystems and the biosphere to result in interconnectedness that we cannot fully take into account in current management practices. This aspect of complexity is one of many that must be taken into account in determining how much water can be used sustainably.

Plate 1.2 Grant's gazelle (*Gazella granti*) is a species characterized as a primary consumer, and has an adult body mass of about 55 kg. It exhibits a measurable population density, population variation, and home range size; it has a life history strategy involving its own characteristic birth rate, mortality rate, and mean age at first reproduction. Relationships among such characteristics result in patterns scientists discover and explain.

Plate 2.1 Many species are migratory. Exemplified by the sandhill crane (*Grus canadensis*), these species underline the reality of interactions among ecosystems. Such species spend one part of their lives in one ecosystem and other parts of their life cycle in other ecosystems. What happens in one of these ecosystems influences what happens in the others—interconnectedness.

Plate 2.2 The behavior of various species is part of what lends to their influence on other species and how other species influence them. The elk (*Cervus canadensis*) exhibits mating behavior, as do many species, that may serve to draw the attention of predators. This can happen through displays that involve color, motion, sound, or odors. Predators can take advantage of the concentration of individuals during times when they are gathered together for reproduction. Such concentrations also intensify the consumption of resources in, or near, such locations.

Plate 3.1 The redheaded woodpecker (*Melanerpes erythrocephalus*) exemplifies a species as a biological system emergent as a product of all contributing factors as depicted in Figure 1.4. Both the individuals, the species, and the ecosystems of which they are a part are such systems. This species is a part of various groups of species, each of which exhibit patterns that are also emergent from the complexity of contributing factors. The genomes of such species represent information about that complexity (Photo courtesy of, and copyright by, Bruce Fowler).

Plate 3.2 Poppies (*Papaver spp.*) are represented by over one hundred species and, as a group, the challenge of taxonomy in identifying just what constitutes a species. Individual species often have their names changed; two or more species are often lumped, later to be split again. Inconsistency in common names presents even more frustration as is experienced repeatedly by bird enthusiasts.

Plate 4.1 Insects such as this cicada are often considered lesser than humans. Such human value systems are brought to decision making in conventional management. In a fully ecosystem-based approach to management all species must be considered and given consideration without bias—either anthropocentric or biocentric.

Plate 4.2 Predator control has been a major component of conventional management. This kind of mis-directed reductionism is not unique in today's world. Any research showing that lions compete with more popular species is verification of competition, not basis for reducing lion populations.

Plate 5.1 Systemic management accounts for factors such as the evolution of thermoregulation (exemplified by large surface areas of body parts such as the ears of the jack rabbit, *Lepus townsendii*) in management based on consonant patterns. This happens without explicit information because of the contribution of such factors to any pattern and explicitly when factors such as ear size are considered in correlative relationships within patterns (Photo courtesy of, and copyright by, Bruce Fowler).

Plate 5.2 The ghost crab, *Ocypode gaudichaudii*, typifies species for which the tide is an important factor. The tide, in turn, is influenced by the gravitational forces of the sun and moon. The latitudinal differences in the influence of the sun are also well known in the amount of energy that passes through primary producers and on to primary consumers—all factors that contribute to patterns among species and, as such, are factors that get taken into account in the use of patterns to guide management.

Plate 6.1 The common milkweed (*Asclepias syriaca*) is a species for which the energy in air currents provide a means of distribution. This is not energy used to achieve a conscious purpose. Evolution has led to success among plants and animals that have seeds or larvae small enough to be transported by air and water currents—among the many effects of the physical environment.

Plate 6.2 The manatee (*Trichechus manatus*) is an endangered species with a population size reflective of such things as its specialized habitat, diet, and body size. Among the many things involved in determining the population size of this species are the cumulative effects of human impacts—thus, factors reflected in the population size we observe for this species today. The pattern in population size among mammalian species with the manatee's body size accounts for human impact.

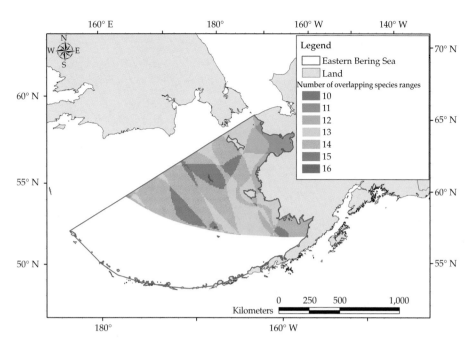

Plate 6.3 Map of the geographic range of the bearded seal (*Erignathus barbatus*, multicolored area) as it overlaps with the eastern Bering Sea ecosystem (i.e., not its full species-wide geographic range). The color coding of this area shows the number of other marine mammal species (N = 20) that have geographic ranges that overlap with that of the bearded seal, each one overlapping different portions of this part of the bearded seal's full geographic range. Thus, the distribution of the colors shows the spatial distribution of these overlapping areas in terms of numbers of species that occupy that area at some time or another within each year (from Fowler and Johnson in prep.).

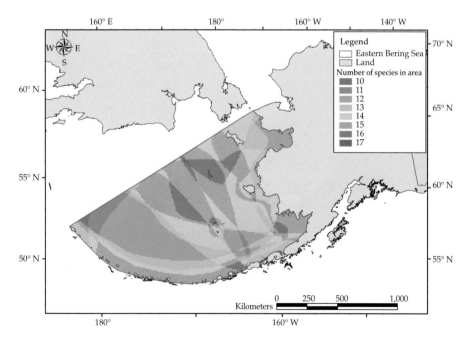

Plate 6.4 Map of the eastern Bering Sea showing the density of marine mammal species found by counting the species with geographic ranges that overlap each point in the system (N = 21, from Fowler and Johnson in prep.). This is the same kind of map as shown for the bearded seal (Plate 6.3) but here involves the entire eastern Bering Sea to include the bearded seal in the total count.

Plate 7.1 Anyone can notice change, and change can include the decline or increase in the size of a glacier anywhere. If there is the least suspicion by anyone that a receding glacier might be related to our production of CO_2, directly or indirectly, there are grounds for looking at the pattern in CO_2 production among other species to see what is normal.

Plate 7.2 Over the eons of time, history has left behind clues as to what happened. Some are found in rock forced up from the bottoms of ancient lakes and seabeds by plate tectonics. We know that everything has a history—everything has an explanation. Everything reflects the complete set of factors behind its emergence. This sacred quality of what we see involves seeing patterns that show us what works.

bottom panel) to achieve levels corresponding to the statistical central tendencies observed in predation rates. For the application of systemic management, this figure conveys both an initial impression of the concept and, at the same time, a rough indication of the dramatic challenges before us. Further examples in the sections below will provide more detail regarding both the concept and the challenges. In Chapter 5, standards of reference other than statistical metrics were introduced (e.g., maximizing biodiversity) and will be revisited below.

Current management of the harvest of walleye pollock is based on conventional single-species approaches. Estimates of biomass, recruitment, and production are converted to advice for management of harvest rates (lacking, avoiding, or rejecting empirical information consonant with the management questions; top row Fig. 1.1). To involve ecosystem considerations, harvests are reduced, in conventional management, from what would otherwise be considered reasonable, by an *ad hoc* and incomplete consideration (by teams of advisors, scientists, stakeholders) of only a few other factors, such as the needs of other predators for walleye pollock. For example, such reductions would include smaller allowable takes in consideration of an endangered species for which pollock serves as a prey species. Takes would be reduced from conventionally established harvest levels based on maximum sustainable yield (MSY) approaches, or spawner-recruit relationships, and models of a limited number of factors only partially related to the management question being addressed. The evolutionary pressures introduced by such harvest rates and their direct genetic effects are unknown and are not taken into account. Also unknown and unaccounted for are the effects of the harvest in all other interactions among other associated species in the ecosystem, including indirect evolutionary effects (i.e., higher order coevolutionary interactions). Most of the unintended consequences of the resulting harvest rates are not taken into account. Among the questions left unasked is: "How much pollock should be left for all other species, the ecosystem, and the biosphere?" Socioeconomic and other short-term human values count heavily among factors used to decide what harvest rates

should be. In an era of management in which the burden of proof is increasingly placed on the users (Holling and Meffe 1996, Mangel *et al.* 1996, Wood 1994), taking more than the extremes (e.g., above the 0.95 statistical confidence limits) within the normal ranges of natural variation among other species cannot be justified scientifically and is in violation of Management Tenet 5. Official certification of such an approach as sustainable is also in violation of Management Tenet 5.

Another initial application of systemic management would be to use the mode[4] in a pattern such as that shown in Figure 6.1 as a first approximation of an estimate of the most risk free and sustainable harvest level achievable. So far, no single statistical measure has been identified as best, but a harvest rate near to central tendencies of distributions like those illustrated above would be better than what is practiced in current management. Finding a harvest level that maximizes biodiversity/information is probably a preferred option (Fig. 5.3, Fowler 2008). As will be seen later, finding ourselves within the confidence limits of such variation is no guarantee that we are sustainable, but would present less justification for change than in cases where we are clearly abnormal.

Systemic management is not a matter of avoiding abnormality in only a few ways. For example, the consumption (harvest) of biomass is only one aspect of resource use. We harvest both biomass and individual fish when we harvest walleye pollock. What is the most sustainable harvest of walleye pollock measured as numbers of fish? A totally different appraisal emerges if the information brought to bear is the numbers of individual fish consumed rather than biomass (Fig. 6.3, especially the bottom panel). Recent takes by humans, measured in numbers, is below the mean for predation rates among nonhuman predators, but well above the mode (linear scale). The difference between Figures 6.1 and 6.3 is explained by the fact that current management focuses more on harvesting adult fish than younger fish. Such selectivity will be dealt with below. The important point to be made here, is that what appears on the surface to be sustainable based on numbers alone can be erroneously interpreted to indicate that current management is sustainable. However, such is clearly *not* the case in terms of biomass (Fig. 6.1). Numbers and biomass are two different dimensions and therefore two different aspects of the complexity of managing our harvest of a resource species; they involve two different management questions. Were we to offer management advice by using the mean of the data for numbers in the data for Figure 6.3 alone we would run the risk of violating sustainability in our harvest of biomass.[5] This begins to provide insight into the complexity of systemic management, a point yet to be fully developed.

Thus, even in the case of numbers alone, things are more complicated than they seem at first glance. In using Figure 6.3 we have not explicitly accounted for allocation of harvests across age. Taking as many adult walleye pollock as are taken in commercial fisheries may not be sustainable owing to both abnormal harvest rates and size selectivity. The size (and, therefore, genetic) selectivity of commercial fisheries does not match that of the nonhuman predators (Etnier and Fowler 2005). Below, we will treat this issue directly.

Figure 6.4 shows the take from several other individual species of fish by commercial fisheries in comparison to that of other predatory species (from Overholtz *et al.* 1991)[6] exactly as was done for walleye pollock above. In each case, the consumption by humans of each of the three species is above the central tendencies, and it is outside (or clearly at the edge of) the normal range of natural variability for two (hake and herring). Spiny dogfish are the only species with consumption rates larger than that for humans in the consumption of mackerel. This species is one of the elasmobranchs that often predominate in parts of the northwest Atlantic as a result of human influence (Sherman 1994). This points out the need to avoid the dangers of assessing and guiding human involvement in ecosystems through comparison with any single species and emphasizes the need to account for abnormal human influence (which is accomplished systemically in these data).

The limitations that would be imposed on human consumption through systemic management of the fisheries depicted in Figure 6.4 would have implications for the fishing industry similar to those for the walleye pollock fishery (Fig. 6.2). For herring, take by humans should have been about 2% (98% less)

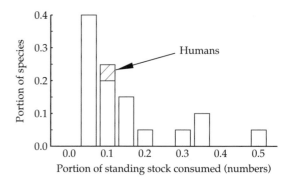

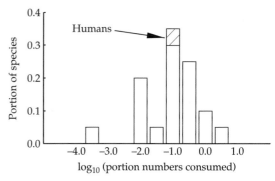

Figure 6.3 Frequency distribution of 21 vertebrate species, including humans, that consume walleye pollock (*Theragra chalcogramma*) in terms of fraction of numbers of individual walleye pollock consumed (data are shown in linear scale in the top panel and in $\log_{10}$ scale in the bottom panel; from Livingston 1993 and personal communication). Humans are represented by the respective cross hatched portions of the bar corresponding to the rate of harvest taken by commercial fisheries.

of that reported for 1988–1992 in order to correspond to the mean consumption rates for other predators. Hake harvests should have been about 10%, and mackerel harvests should have been less than 20% of the harvest rates of the same period in order to match the mean consumption rates of other predators on the same species. If these data were from systems undisturbed by abnormal human influence, they would reflect rough approximations of ultimate sustainability for these systems. As they are, they clearly represent systems subject to what was abnormal human influence. Based on these comparisons alone, we are still several steps away from understanding the full application of systemic management. For example, are the means of such distributions the best measures of sustainability or

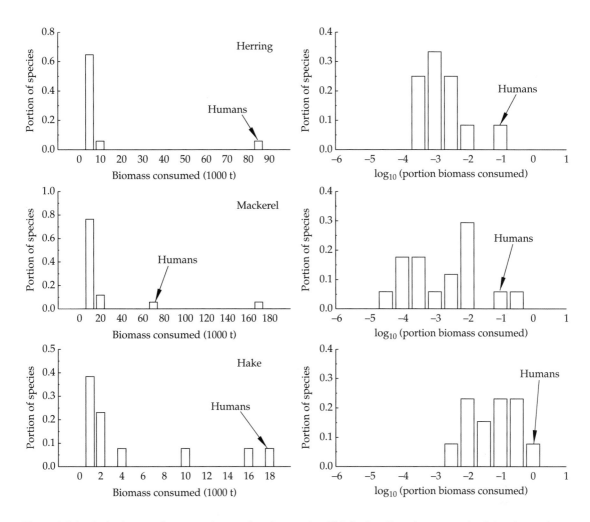

Figure 6.4 Species-level patterns for consumption rates from three species of fish (herring: *Clupea harengus*, mackerel: *Scomber scombrus*, and hake: *Merluccius bilinearis*) serving as resources in northwest Atlantic fisheries as modified from Appendix Figure 2.1.5 to show the harvest rates by humans by comparison. Total biomass consumed is shown in the left column and the corresponding portion of standing stock biomass consumed is shown in the right column which is presented in $\log_{10}$ scale, from Overholtz *et al.* (1991).

are the predatory species represented by these data the best examples of sustainability for our species? There are more such questions—all related to the application of systemic management. How do we account for the effects of functional responses of predators to prey abundance? What are the ways in which abnormal human influence can be taken into account more directly, that is, other than knowing that these patterns reflect human influence?

The harvest of resources is not confined to fish, or marine ecosystems. The sport harvest of

ungulates provides contrast, both in being within terrestrial ecosystems, and in being an example wherein humans are not abnormal compared to other species. Figure 6.5 shows mortality rates by nonhuman species in their predation on three species of ungulates (females of *Odocoileus virginianus*, *Cervus elaphus*, and *Alces alces*) in a specific region of the United States and Canada (Kunkel and Pletscher 1999). On the surface, there is no reason to be concerned about the harvest rates used in this system, measured in simple terms of

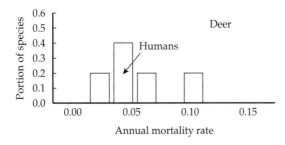

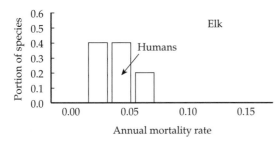

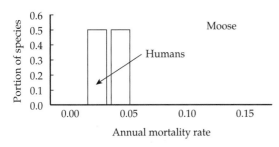

Figure 6.5 The variability in mortality rates observed in predation on three species of ungulates by their predators in the Forth Fork of the Flathead Valley of northwestern Montana and southeastern British Columbia (Kunkel and Pletscher 1999). The data represent four species of nonhuman predators for deer (*Odocoileus virginianus*) and elk (*Cervus elaphus*), and three for moose (*Alces alces*).

the pattern in mortality rates caused. As with the fishery examples above, however, management also involves decisions about selectivity by age, size, and other genotypic characteristics. There are other management questions; the sustainability of harvests of males needs to be addressed with patterns in selectivity by sex.

The refinement of management questions, in this case, goes beyond the single question "What

is a sustainable harvest of deer?" Other questions need to be addressed in attempts to tackle the complexity of nature. These include: "What is a sustainable harvest for species with the body size and reproductive rate of deer when taken by species with the body size of humans?" In this example there is little concern about consonance in regard to body size as the predators (*Canus lupus, Puma concolor, Ursus arctos, Ursus americanus, Canus latrans*) are close to human body size, and are mammals. Concerns that are left to be addressed involve issues such as genetic selectivity (which generates new questions).

The research by Kunkel and Pletscher (1999) exemplifies studies that succeed in producing information involving patterns that are consonant with management questions regarding sustainable harvest rates—in this case, questions about sustainable harvest rates of ungulates.

The examples for fish and ungulates provide an introduction to the concept of systemic management applied in our interaction and influence on individual species. They offer an assessment of our species-level roles in ecosystems by comparing humans with other species in species-to-species interactions. They provide crude indications of the changes needed in fisheries management to satisfy the tenets of management (Chapters 1 and 5). In management action (praxis), humans are made subject to control (i.e., through action wherein control is more of an option by limiting harvests, Management Tenet 2). Furthermore, these examples involve our interactions, as a species, with other *individual* species. It is thus a matter of single species application of systemic management regardless of how it is viewed. Through such examples, we begin to account for complexity because ecosystems, other more inclusive systems, and history are all reflected in the patterns among species used as sources of guiding information (Fig. 1.4; Belgrano and Fowler 2008, Fowler 2002, Fowler and Perez 1999, Fowler *et al.* 1999, Fowler and Hobbs 2002, Fowler and Crawford 2004; Appendix 4.4). This complexity contributes to the factors that prevent the accumulation of species that might have otherwise fallen outside the normal range of natural variation. Thus, complexity is accounted for in the guiding information (Fig. 1.4) as an indirect process. However, it must be clear that deciding to take our harvests so that they fall within the normal ranges of natural variation for consumption levels observed among nonhuman predators alone, is just a beginning and, with the data we have in hand, we only begin to see evidence of precision in guidance. We have yet to achieve a complete understanding of how to overtly or directly account for a variety of factors integral to the patterns.

6.1.1.1.1 Dealing with abnormal anthropogenic effects, past and present

Figure 6.6 illustrates changes that occurred in the biomass of cetacean populations in the eastern Bering Sea, at least partially due to commercial whaling. Thus, estimates of the biomass consumed by cetacean species also differ between these two periods. The total biomass of cetaceans in recent years has been about one third of what it was in the 1940s. This is very close to the change in the consumption rates observed between these two periods of time as shown in Figure 6.6. Likewise, estimates of biomass consumed by all nonhuman mammalian species are less now than before recent anthropogenic effects. Populations of fur seals (*Callorhinus ursinus*), northern sea lions (*Eumetopias jubatus*), and other species are below historically observed levels (NRC 1996). Therefore, we likely would find that advisable sustainable harvest levels to be taken from walleye pollock would be higher if based on data for previous periods (i.e., from ecosystems free of abnormal disturbance by humans) than is indicated in Figures 6.1 and 6.2 (e.g., three times as much if the data in Fig. 6.6 are at all representative). The data behind Figures 6.1 and 6.2 are based on an ecosystem already modified by the various abnormal effects of humans—effects that are outside the normal range of natural variation (including harvesting, CO_2 production, and pollution).

Although current data indicate what is currently sustainable (i.e., account for human impacts insofar as they involve reactions to human influence), at some point in the future, we will need to have some idea of what is ultimately sustainable (with humans fitting in normally in all ways conceivable). Such estimates can be achieved in at least three ways, a combination of which is advisable. First, we can use

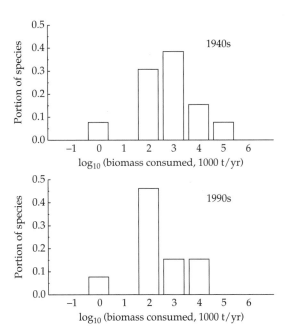

Figure 6.6 Annual biomass consumption from the Bering Sea by cetacean populations as estimated by Sobolevsky and Mathiesen (1996) for the 1940s and 1990s.

data from historical monitoring (or estimates of historical conditions) such as would be the case in estimating the consumption rates for the cetaceans in the top panel of Figure 6.6. These were presumably subject to less abnormal human influence than are the systems observed today. Second, we can monitor the consumption by other species as the system recovers from human disturbance in response to management that places humans within the normal range of natural variation for all of the ways we are now abnormal (e.g., confining human consumption to levels within the normal range of natural variation, reducing CO_2 production, finding normal genetic selectivity in our harvests—all the ways normality can be achieved). Third, we can use information from comparable systems at other times and places in correlative relationships.

6.1.1.1.2 Dealing with correlated species-level characteristics

Medical doctors do not give the same advice to men and women, or to adults and children, for managing diets and exercise to obtain advisable body weight. Similarly, we treat neither the heart rate nor the ingestion of food the same for shrews and elephants; they are different species with different characteristics. This is true for individual shrews and elephants and for each individual species. Body size, trophic level, and metabolism are among the factors to take into account. We are a homeothermic mammalian species with an average body size of approximately 60 to 70 kg. Thus, small heterothermic fish species are not as appropriate as examples of sustainability for humans as other mammalian species would be. Figure 6.7 illustrates the frequency distributions that would be used in systemic management of our harvests in the single-species approaches introduced above if such management were based on data from other

mammals. Our take of walleye pollock would be reduced to 3.6% of recent takes if we assume that using the mean among mammalian examples suffices to provide adequate guiding information. For hake, herring, and mackerel, the reductions would be to 7.0%, 3.3%, and 5.9% of recent harvests, respectively. As can be seen, the more complexity we deal with, and, in particular, when various factors are dealt with directly or overtly, the more we appreciate the degree to which we have failed to account for complexity, and failed to achieve sustainability in current management. Although using data from other mammals is critically necessary, by itself, it is insufficient for dealing with complexity completely. For example, Figures 6.6 and 6.7 do not overtly account for body size appropriate for humans.

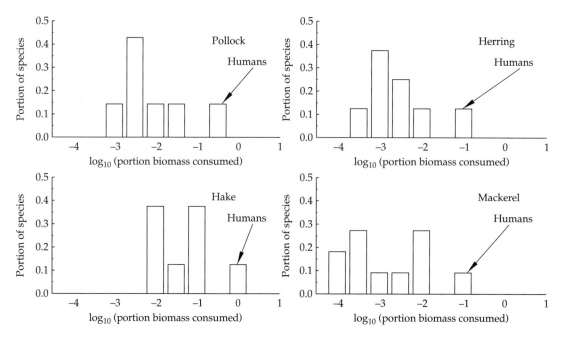

Figure 6.7 The consumption (harvest) of fish by commercial fisheries compared to that by marine mammals for four species of fish prey. Comparison is illustrated in frequency distributions for rates measured as $\log_{10}$ of the portion of the standing stock biomass of prey taken annually. Commercial takes of walleye pollock are compared to consumption by six other mammalian species (e.g., northern fur seals, *Callorhinus ursinus*, the bar to the left of the fisheries take, and harbor seals, *Phoca vitulina*, one of the three species represented by the tallest bar). Takes of herring by fisheries are compared to consumption by seven species of mammals (e.g., fin whales, *Balaenoptera physalus* in the uppermost bar below humans). The consumption of hake involves a comparison of commercial fishing with seven species of marine mammals, including harbor porpoise (*Phocoena phocoena*, represented by the shortest bar among the marine mammals). The consumption of mackerel is represented by ten species of marine mammals (e.g., minke whales, *Balaenoptera acutorostrata*, one of two species in the bar at the far left). The data presented here represent the same systems shown in Figures 6.1 and 6.4.

6.1.1.1.3 Dealing with environmental factors

The evolutionary history of species, including environmental factors, and all human influence, are part of the explanation of natural patterns such that they represent an integration of such factors (Fig. 1.4). Because species in different environments respond to different seasonal and decadal changes, the distributions for the same species may be different under different circumstances. Thus, to further account for complexity directly, these dynamics must be taken into account in the same way species-level characteristics were taken into account above. The dynamics of species frequency distributions over time, differences over space, and reactions to environmental circumstances can be used as information. Patterns change with circumstances that can be treated explicitly. Thus, we could use distributions like those presented in Figure 6.7 for the harvest of individual species during years of climate corresponding to that during which the data were collected. We would need different sets of data for alternate climate regimes and different population levels of the resource species (while, of course, simultaneously accounting for other species-level characteristics as discussed above).

6.1.1.2 Multispecies approaches

What is a sustainable harvest from a particular group of fish species? Here we are expanding the application of systemic management from the single-species approaches introduced above. We are now considering human interactions with groups of species in the progression toward interactions with ecosystems and the biosphere. Communities, guilds, taxonomic groups, or trophic assemblies of species are also subject to human effects that need regulation. How much of the production by any species, or group of species should be left unharvested to sustain such systems? Each needs resources and energy to sustain its normal structure and function. Groups can be defined in numerous ways, including spatially determined groups, groups of specific body size, groups of specified life-history traits, groups with specific reproductive strategies, groups preyed upon by a specific predator, or groups defined by numbers of predator species that prey upon them. A group of particular interest to

management involves predator/prey pairs, always two in the simplest case, but which also could include a predator and all of its prey. Management is required to consider such groups, all as parts of the spectrum from individual species to the biosphere, and all in accounting for complexity at various levels of biological organization. Finding a way to regulate our interactions with such groups is thus a piece of systemic management; the parts of systemic management include all such interactions. Resource use serves as a continuing theme in demonstrating how to take advantage of information on species-level patterns, but again is not the only element of human influence to be involved in the full treatment of complexity.

The limits to natural variation in resource consumption from several groups of species are demonstrated in Figure 6.8. This figure shows consumption rates by humans in comparison to the natural variation in the consumption rates among consumers feeding on three different groups of fish. First, the finfish of the eastern Bering Sea are a taxonomic group (panel A, Fig. 6.8) fed upon by 20 species of marine mammals. Consumption by humans would have to be reduced by about 97% to correspond to the mean of consumption rates by nonhuman species based on these data. Second, mackerel, herring, and hake are commercially valuable fish in the northwest Atlantic (panel B, Fig. 6.8). The total harvest of these species would have to be reduced from recent harvest levels by about 98% to correspond to the mean of the consumption rates estimated for the marine mammals (about 3000 metric tons per year). Third, the fish off the southwest coast of Africa include four species, some of which are of commercial value. If all were harvested commercially, the data represented in panel C of Figure 6.8 would imply a need to reduce the overall take of biomass from this collection by more than 99% of recent harvests (primarily catches of anchovy at the rate of about 480,000 metric tons per year), continuing to assume that we have adequately matched (achieved consonance[7] between) empirical pattern and management question (a failure we obviously risk in comparing ourselves to birds).

What is a sustainable harvest from the combination of deer, elk, and moose in the system depicted

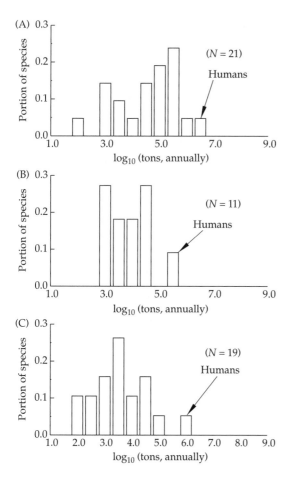

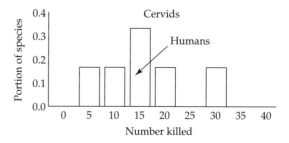

Figure 6.9 The pattern of variation in total number of female animals killed by the predators and human hunters in the cervid species-group shown as individual species in Fig. 6.5. Whereas the rates shown in Fig. 6.5 apply to the respective populations, the numbers shown here are pertinent to the population of the species group defined by the animals tagged in the study.

Figure 6.8 Variation in consumption rates (log$_{10}$ metric tons of biomass consumed annually), comparing commercial harvest by humans with consumption by other predators: (A) human and marine mammal predation on finfish in the eastern Bering Sea (Fowler and Perez 1999); (B) human and marine mammal predation on hake, herring, and mackerel in the northwest Atlantic (same species as represented in Figs. 6.4 and 7.7; Overholtz *et al.* 1991, and personal communication); (C) human and seabird predation on lanternfish (*Lampanyctodes hectoris*), lightfish (*Mourolicus muelleri*), and anchovy (*Engraulis capensis*) off the southwest coast of Africa (Crawford *et al.* 1991). Numbers in parentheses are samples sizes (numbers of species) including humans. See original sources for lists of predators involved.

in Figure 6.5? Figure 6.9 shows the pattern consonant with this management question when measured in units of numbers of individual animals. Not only does this demonstrate a multispecies application in a terrestrial system, it again shows

humans to be well within the limits of variation—again, not necessarily fully sustainable.

What is an advisable allocation of take over the three resource species in the ungulate example? Deer made up about 36% of the cervids harvest by humans compared to 0–100% for nonhuman predators (mean of about 48%). Elk made up about 43% of the total cervid harvest by humans compared to 0–46% for the nonhuman species (mean of 28%). Moose made up about 24% of the cervid harvest by humans compared to 0–42% (mean of 24%) for nonhuman predators. As with the harvest of individual species, the allocation of harvests by humans across the individual species show no abnormality compared to the observed variation for the nonhuman consumer species (although the harvest of elk might have been close to an excessively large portion of the overall harvest).

What portion of the production by deer, elk, and moose should be left for the other species within the ecosystem (and for normal recruitment to the respective populations)? These and similar questions can be addressed consistently for both fisheries and ungulate systems (Fowler and Hobbs 2008).

6.1.1.3 Ecosystem approaches

We would be guilty of repeating past errors if we believed that managing our interactions with specific nonhuman species, species groups, or the combination of individual species and species groups is sufficient, even by avoiding the abnormal in every case. As we understand from the principles of management, complexity requires considering all components of reality (Management Tenets 3 and 4, Chapter 1). Other questions remain: How many species can be harvested if we harvest them individually as introduced above? What is the total harvest that can be taken sustainably from any particular ecosystem? What portions of the production and standing stock of species groups, ecosystems, and the biosphere should not be harvested? How do we allocate our harvests among alternative species? How many trophic levels, and what range of trophic levels, should be considered in our selection of resource species? How should we allocate our harvest among trophic levels? The patterns of natural systems provide answers to these questions.

For example, Figure 6.6 illustrates the concept of resource use within ecosystems by showing the annual take of biomass from the Bering Sea by cetaceans. Management of our use of ecosystems can use such information as preliminary guidance based on the observed limits to natural variation. There are two ways to account for some of the influence of humans that has been historically (especially recently) outside the normal range of natural variation. First, the top panel of Figure 6.6 is a better indicator than the bottom panel of what ultimately can be sustainably taken from the Bering Sea after it has been allowed to recover from abnormal past and current anthropogenic influences. Second, the bottom panel is more of an indication of what is sustainable under current circumstances to account for the plethora of current and past human influences. Many cetaceans, however, are larger than humans and we need to account for factors such as our body size as directly as possible.

Figure 6.10 shows preliminary patterns in sustainable harvest rates in consumption of biomass from the eastern Bering Sea, the Georges Bank, and the Benguela ecosystems. Figure 6.11 shows the change that would have to be made in harvests from the eastern Bering Sea if we were to use the mean of the data from Figure 6.10 (top panel) as a management goal. The example in Figure 6.11

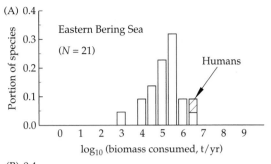

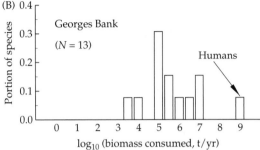

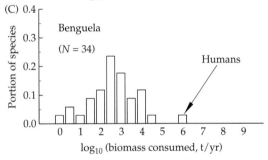

Figure 6.10 Variation in consumption rates ($\log_{10}$ metric tons of biomass consumed annually), comparing commercial harvest by humans with consumption by other predators for three ecosystems: (A) humans and marine mammals in the eastern Bering Sea ecosystem north of the Aleutian Islands (Fowler and Perez 1999), (B) humans and marine mammals in the Georges Bank ecosystem of the northwest Atlantic (Backus and Bourne 1986), and (C) humans and seabirds in the Benguela ecosystem off the southwest coast of Africa (Crawford *et al.* 1991). Sample sizes shown in parentheses (number of species) include humans.

again emphasizes the importance of assumptions being made. Smoothed symmetric curves cannot adequately represent the underlying data, the geometric mean is probably not the best standard, the pattern in hand may not apply in today's world, and there are undoubtedly other correlative factors that can be brought to bear in further analysis.

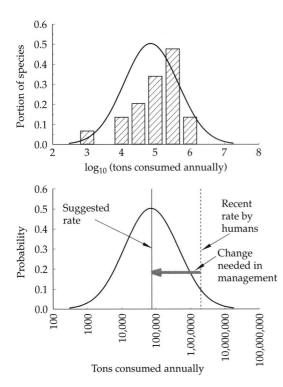

Figure 6.11 Illustration of the change needed for systemic management to bring human consumption into the normal range of natural variation (in this case, equivalent to the mean) of the total harvest of biomass from the ecosystem of the eastern Bering Sea, based on data for consumption rates by 20 species of marine mammals.

However, as a preliminary indication of needed change based on these data, the total harvest of biomass from the eastern Bering Sea would have to be reduced by about 97% (from about 2 million metric tons per year, late in the 20th century, to about 60,000) to correspond to the mean of biomass consumption by mammalian species.

We need to further deal with complexity through the use of correlative information in the management of our use of ecosystems, just as we do for our use of individual species or groups of species. Again, this means accounting for anthropogenic influence, correlated species-level characteristics, and environmental circumstances. Data based on consumption among marine mammals prior to abnormal human influence by humans, would be

better than what we now have in hand, for estimating what might be sustainable after ecosystems have recovered from abnormal human influence. Data for the consumption rates among species of a body size similar to that of humans serves better than data for species of other body size or taxa. Thus, data for marine mammals would be preferable to that for marine birds in the Benguela ecosystem (marine mammals are not included in the bottom panel of Fig. 6.10). Data on consumption collected during climatic conditions similar to the present would add to the quality of the resulting guidance. Further accuracy would be achieved with systems free of abnormal human influence long enough to have recovered.

6.1.1.4 Biosphere approaches, marine and terrestrial environments

Hierarchical complexity extends beyond species, species groups, or ecosystems to the biosphere. In each case we currently have very limited data for explicitly considering human characteristics and environmental factors in correlative relationships. On the conceptual path from species to the biosphere, however, we pass through the marine environment. The fractal-like pattern of systemic management becomes more apparent as we proceed through the various levels of biological organization. We continue with the theme of resource use in our introduction of information to guide systemic management, being mindful that many other management questions need to be addressed.

Figure 6.12 compares consumption of biomass by humans to the pattern among other species. It shows limits to natural variation seen in the consumption of biomass from both marine and terrestrial environments, as well as that for the biosphere, based on information for consumption by various mammalian consumers. Managing to make the removal of biomass from the marine environment in fisheries (i.e., consumption by humans) correspond to the mean among other species of mammals would require a reduction of about 99%. To make our ingestion of biomass in the biosphere correspond to the mean of that among other mammalian species would require a reduction of over 99.9%. We now see, with growing clarity, the magnitude of our problems, which, without solution, will

a major factor in creating problems identified for other species, ecosystems, and the biosphere as identified in earlier chapters (e.g., Appendix 4.2). Determining all of the effects on ecosystems of both obtaining and using energy is an impossible task but involves numerous detailed questions for which we can find consonant information (e.g., amount of nonbiodegradable materials produced, patterns in surface area covered by structures, amount of soil moved, energy used in food storage/refrigeration, and average distance traveled in a lifetime). Various aspects of this influence include harvesting food as an energy source for metabolic needs, and wood for fuel; each represents extraction from the system with corresponding ecosystem-level effects. The energy used in these activities also has effects, including those activities that do not involve the extraction of energy for ingestion (e.g., construction, manufacturing, production, distribution, waste disposal, etc.). See later sections of this chapter for consideration of factors involved in extinction.

Another approach to evaluating energy usage is to employ an index represented by the overall fraction of ecosystem-level energy budgets that come under human influence. Such an index of effects would be a measure that can be compared among species in a way that applies to the entire biosphere. For example, Vitousek *et al.* (1986) and Wright (1990) have indicated that humans now use, influence, or control between 20% and 40% of the Earth's primary productivity. This is almost six orders of magnitude greater than the mean of the same influence by other species (Fig. 6.18, Fowler 2008).[12] Management to remove the abnormality of human energy use shown in Figures 6.15 and 6.18 would dramatically reduce human influence through energy consumption whether it be on any species (including our own), full scale ecosystems, or the biosphere. The caveats noted for earlier examples also apply here. Systemic management would be based on considering anthropogenic influence on the position of the species in such distributions, other species-level attributes, and environmental circumstances, just as for resource use and CO_2 production (always implicitly, based on the emergence of such patterns, and explicitly, through correlative analysis, as often as possible).

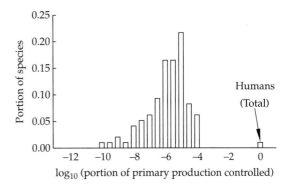

Figure 6.18 Approximate frequency distribution comparing total energy garnered within the biosphere by humans with energy consumed by 96 species of mammals, expressed as a fraction of the energy (2.4×10^{21} joules per year) estimated to be available for higher trophic levels through primary production (Wright 1990).

6.1.5 Geographic range and mobility

Another element of complexity, or another way species can be measured, involves their geographic range, or the geographic space occupied. What portion of an ecosystem, a continent, or the biosphere should humans occupy in the process of using resources, producing CO_2, or consuming energy? The complimentary question must also be asked: What portion should be put into reserves for protecting other species from direct human influence? Empirical patterns find consistent answers (Hobbs and Fowler 2008).

For purposes of appraisal, the current geographic range of our species can be assumed to be 70% of the Earth's terrestrial surface, excluding Antarctica, although some ascribe our presence to as much as 95% (Pimentel *et al.* 1992). Figure 6.19 compares this 70% with the geographic ranges of 523 species of North American mammals of all body sizes. A more realistic comparison would be with mammals of similar body size, a species-level attribute correlated with geographic range size (see Fig. 2.28). If we were to attempt to reduce our geographic range to the mean of that for other species of our body size, we would need to confine ourselves to about 760,000 km² (a reduction of over 99%).

As exemplified above, and in other examples of systemic management, the interrelationship

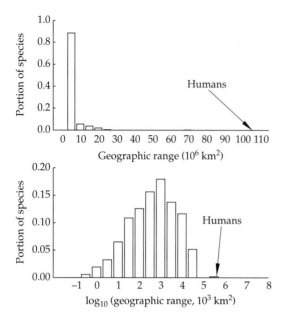

Figure 6.19 Species frequency distributions comparing human geographic range size with 523 mammals of all body sizes found in North America (Pagel *et al.* 1991a and personal communication, 4/30/91). A very small fraction of this sample have geographic ranges that extend beyond North America that are not accounted for in this illustration. The top panel shows actual area, with the geographic range size for humans assumed to be about 70% of the Earth's terrestrial surface, excluding Antarctica. The bottom panel shows the same information in $\log_{10}$ scale.

of species characteristics must be considered. Geographic range cannot be considered to the exclusion of other related factors. Among things accounted for automatically are the facts that as geographic range size increases, more and more ecosystems are influenced. Such increases lead to competing for resources with more and more other species. Increased human involvement in more ecosystems results, among other things, in exposure to more diseases (Garrett 1994).

Mobility is pertinent to the number of ecosystems in which we are involved. Aided by bicycles, cars, trains, and airplanes, many humans travel millions of kilometers per year. It is highly likely that information on the limits to natural variation in distances traveled per unit time would prove humans to be well above the maximum for

other species—certainly in terms of the maximum as measured for individuals of our species who represent the extremes of distance they travel each year. We are undoubtedly exhibiting abnormal mobility in comparison to species of our body size and trophic level. Our mobility affects the mobility of other species (e.g., parasites, intentionally and unintentionally introduced species, and disease organisms) with the related risks.

6.1.6 Population

Few would argue that the changes described in the previous sections can be achieved without resulting in a reduction in the human population. Reducing resource consumption would reduce the human population just as reducing the human population would reduce resource consumption. The aspect of complexity we confront here is that while things are interconnected, solving one problem does not guarantee that we solve all problems. Systemic management takes advantage of connections to identify other management questions—each to be addressed systemically. Here the connection between population and other problems (e.g., CO_2 production, resource consumption, and geographic range size) highlights the need to be sure that they are not ignored. Reversing the connection—noting the reciprocal connection between resource consumption and population size—emphasizes that the population issue must be confronted as directly as possible on its own terms. Every issue must be evaluated directly (i.e., with empirical patterns directly matching, or consonant with, the question regarding what is sustainable). Also, it is important to know that dealing with only a select set of problems can run serious risks. In this section we deal directly with human overpopulation as an issue raised by consideration of the challenges discussed above—abnormal consumption of biomass and energy, abnormal CO_2 production, and geographic range size far outside the normal range of natural variation. As was seen in Chapter 5, the consideration of one issue leads to the discovery of others. All will require changes by each of us individually, as well as socially, if we are to succeed willfully as opposed to being victims of systemic change.

What, then, is a sustainable number of human beings? The following sections illustrate how patterns can be used to estimate a sustainable human population size. The results are compared to estimates in the literature compiled through conventional approaches and the current application of science. We begin with the latter.

6.1.6.1 *Conventional considerations*

In the scientific and popular literature alike, most references to the problem[13] of overpopulation are qualitative. The problem is identified as excessive numbers of humans (Appendix 6.2), but *how* excessive is rarely evaluated (Catton 1980). Existing evaluations are usually made on the basis of conventional methods (see Cohen 1995a,b) and almost always focus on one specific relevant factor such as water, energy, space, or resources rather than the combination of all relevant factors. Factors deemed relevant are usually considered individually but occasionally are treated in small subsets. The full set of factors is never dealt with completely; the complexity of reality is never fully considered. For example, no full accounting for the impact of our water use has been carried out in a way that simultaneously accounts for all other factors such as habitat, energy use, and coevolutionary processes (to meet the requirements of Management Tenets 3, 4, and 7). The impossibility of doing so in conventional approaches would seem to bring us to an impasse. However, the problem is overcome by using existing empirical normative information (Management Tenet 5) directly addressing population size. By doing so, everything is taken into account (Fig. 1.4, Belgrano and Fowler 2008). Cohen (1997), Fowler and Perez (1999), and Fowler (2005) take initial steps in this more systemic form of evaluation.

Conventional approaches to assessing the carrying capacity of the Earth for humans are summarized in Table 6.1 and Figure 6.20. Table 6.1 presents a variety of numerical estimates of sustainable levels of the human population. Figure 6.20 shows a frequency distribution of other estimates compiled by Cohen (1995b, and as explained in Appendix 6.3). The collection of scientific opinion represented in Table 6.1 leads to the conclusion that an optimal human population would be

between 5% and 35% of late 20th century levels, a range of about 0.29 to 1.99 billion. This is different from the mean of estimates in Figure 6.20 (55.02 billion; Cohen 1995b, Appendix 6.3) by approximately an order of magnitude. According to the estimates from Cohen (1995b), the human population could increase almost tenfold to reach this mean and still be sustainable.

The difference between those two sets of assessments is explained by the fact that estimates in Table 6.1 are based on combinations of a few selected ecological factors and relatively complex systems approaches, while estimates in Figure 6.20 are based primarily on consideration of single factors or single species population models for humans.

The human population can also be assessed based on comparisons with historic levels. Today's human population of over 6 billion is as much as 400 times denser and 1000 times larger than 10,000 years ago, before the beginning of organized agriculture, when human population is estimated to have been about 5 to 10 million (see Appendix 6.3). These numbers can and will be debated, a point to be emphasized, but not to the detriment of the quality of difference since there is no doubt that the current population is many times denser and more numerous than in prehistoric periods.

6.1.6.2 *Human population density evaluated with information on empirical limits to natural variation*

Current human population density is greater than any other herbivorous species of our body size (Fig. 6.21). Assuming a human population density of about 11.3 per square kilometer (Appendix 6.3), the population as we enter the 21st century is about 4.9 times the mean expected for herbivores of similar size. This assumes the entire Earth is sustainably inhabitable by humans (Table 6.2). By comparison, the mean density of humans for all terrestrial habitat except Antarctica, is 55.9 per square kilometer, 24.2 times the mean for similar-sized species of herbivorous mammals. The mean density of humans for what we conventionally consider the habitable portion of the Earth (assumed to be 20% of non-Antarctic terrestrial area) is 275 per square km, 120-fold greater. Note that this assumed habitable portion of the Earth is 16 times larger than the arithmetic mean geographic range size for other

Table 6.1 Rough estimates of sustainable levels of the human population (billions) for the entire earth (WP = world population) and the overpopulation factor (OP—number of times by which the current world population has exceeded that value) as extracted from the scientific literature and presented in chronological order*

	WP	OP	Source
1	2.78	2.1	L. Brown (1971) (<250 million for U.S.)
2	6.00	1.0	Commoner (1971)
3	0.56	10.4	Hardin (1971)
4	2.28	2.5	Paddock (1971) (<205 million for U.S.)
5	1.03	5.6	Odum (1972) (5 acres per person)
6	1.99	2.9	Rosenzweig (1974) (based on acreage available and required for population in 1970s—210 million)
7	0.50	11.5	Rosenzweig (1974) (based on water supply)
8	0.40	14.4	Catton (1980)
9	0.57	10.2	Catton (1980)
10	0.02	288.5	Walker (1984) (carnivorous humans in particular habitat type)
11	0.52	11.1	Fearnside (1985, 1990) (estimate of indigenous densities)
12	2.00	2.9	Ehrlich as quoted by Tudge (1989)
13	0.10	57.0	Soulé as quoted by Tudge (1989)
14	10.40	0.6	FAO (see Fearnside 1990)
15	1.38	4.2	Ehrlich and Ehrlich (1990) (extrapolated from Chinese estimate for China)
16	6.00	1.0	Meadows (1991)
17	1.89	3.1	Costanza (1992) (current European standards)
18	0.94	6.1	Costanza (1992) (current U.S. standards)
19	1.50	3.8	Ehrlich and Ehrlich (1992)
20	1.52	3.8	Grant (1992) (using mean of 125–150 for U.S.)
21	1.11	5.2	Pimentel and Pimentel (1992) (should be <100 million in U.S.)
22	2.78	2.1	Werbos (1992)
23	0.67	8.7	Werbos (1992)
24	1.47	3.9	Whelpton (1939) (less than the 1939 population—assumed to be 132 million for the U.S.)
25	2.00	2.9	David Pimentel (Feb. 21, 1994, AAAS meeting, San Francisco, CA).

*Some of these estimates are based on country-specific estimates extrapolated to the world by assuming the density estimated for the country applies to 20.4 million km of land habitable by humans (about 6% of the terrestrial surface of the Earth).

mammalian species, and 120 times larger than the geometric mean (Fig. 6.19).

Indices of human overpopulation shown in Table 6.2 are based on comparison to other species from relationships between population density and body size (Damuth 1987, Peters 1983). The numbers in parentheses are the sizes of the human population (in millions) if they were consistent with the population size at the mean for mammals with body size similar to that of humans (assuming a human adult body mass of 68 kg). The numbers without parentheses are indices representing the factor by which the current human population density exceeds the mean density of other mammals of similar body size. Indices are calculated by dividing the measure of human population density by the mean density of mammalian herbivores or carnivores using statistical procedures identified. Density for humans is the total human population divided by the area indicated.

Figure 6.21 shows the derivation of the information in the second column of the bottom row of Table 6.2. A human population, equivalent to the mean for other herbivores (i.e., assuming we

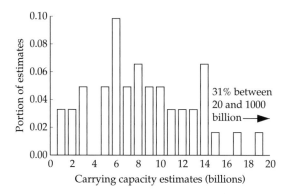

Figure 6.20 The frequency distribution of estimates of the carrying capacity of the Earth for humans based on conventional thinking, primarily single-factor considerations and population models, from the compilation by Cohen (1995b). One estimate of a billion billion is ignored as are those with only an upper or a lower bound.

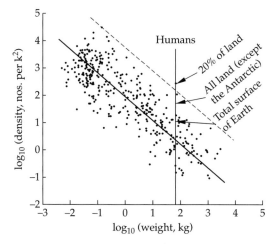

Figure 6.21 The relationship between population density and body size (log transformed) for 368 species of herbivorous mammals from Damuth (1987). The regression line (solid sloped line) is based on geometric mean regression analysis. The dashed line is parallel to the regression line based on (and located above) all data points except that for humans. The vertical line (for human body size) compares human density with that for other mammals of the same size as the human population if it were spread uniformly over varying parts of the Earth (i.e., arrows point to human population density as it depends on the geographic range assumed for humans as identified for each arrow).

are strict vegetarians) of similar body size, would be about 48 million people occupying 20% of the Earth's land surface—a population lower than the Earth has seen for over 6000 years (Catton 1980). (It is to be reiterated that 20% of the Earth's land surface is well above the mean of geographic range sizes seen in Figure 6.19, and about 40 times larger than the mean geographic range size for other species of mammals of our body size.) With a current population about 120 times larger than 48 million, humans are well outside the 95% confidence limits for density among the nonhuman mammalian herbivore species (Fig. 6.22).

Measures of overpopulation increase when we take trophic level into account explicitly as one of the correlative variables behind observed variation. We are not strict herbivores and we need to account for correlative relationships involving trophic level as outlined above in regard to resource use (part of accounting for correlative factors in general). When we do so, we find that the current human population is between 578 and 2470 times more densely populated than the mean of other species of our body size at higher trophic levels (Table 6.2, Appendix 6.4, see also Appendix 6.5 for densities associated with the various nations of the world). Optimal levels for humans may occur between the estimated mean densities for herbivores and those for carnivores in view of our omnivorous habits—direct omnivore information would be of optimal use in this regard. It is entirely possible, however, that mean density for omnivores is higher than either more strict herbivores or carnivores owing to a broader spectrum of species in the diet. Direct (consonant) measures are needed.

Figure 6.23 (A) shows conventional estimates of sustainable human population, many of which use procedures which make direct use of density (Appendix 6.4). In these projections, sustainable density is calculated on the basis of numerous assumptions (e.g., the availability of resources or energy). Density is often combined with assumptions about geographic range (and thus involve the model: $P = A \times D$; where P = population, A = Area occupied, and D = density). Figure 6.23 (B, C) uses empirical information about density (and the correlative pattern involving geographic range size) rather than a select few assumptions

Table 6.2 The human population (millions) evaluated with information for other mammalian species of similar body size showing both expected human population size (in parentheses) and indices of overpopulation (no parentheses) for the indicated areas, trophic levels, and density–body size relationships

Human density determined for	Damuth[1] Herbivores	Damuth[2] Herbivores	Peters[3] Herbivores	Peters[4] General	Peters[5] Carnivores
Total Earth's surface	2.8 (2046)	4.9 (1180)	5.6 (1038)	23.6 (245)	100.7 (57)
Land surface only (excluding Antarctica)	13.8 (417)	24.0 (241)	27.3 (212)	115.6 (50)	494.0 (12)
Twenty percent of land surface inhabitable	69.1 (83)	119.9 (48)	136.3 (42)	577.9 (10)	2470.2 (2)

For D = density in individuals per square kilometer and W = body size in kilograms, the relationships used are
[1] $D = 95.5\ W^{-0.75}$ from Damuth (1981) using mammalian herbivores with ordinary least squares regression.
[2] $D = 91.05\ W^{-0.87}$ using data from Damuth (1987) for mammalian herbivores with geometric mean regression.
[3] $D = 103\ W^{-0.93}$ from Peters (1983) for mammalian herbivores using ordinary least squares regression (see also Damuth 1987).
[4] $D = 30\ W^{-0.98}$ from Peters (1983) for animals in general using ordinary least squares regression (see also Damuth 1987).
[5] $D = 15\ W^{-1.16}$ from Peters (1983) for mammalian carnivores using ordinary least squares regression.

about factors that determine density. Just as harvesting a resource species involves both biomass and numbers, however, populations involve more than density and more than density as related to geographic range size. Dealing with complexity means considering total population size measured directly—consonance with the question being addressed: "What is the most sustainable total population size for humans?" This involves using patterns for total population size to account for complexity, then using those factors selected for conventional consideration in correlative analysis of such patterns.

6.1.6.3 Total population evaluated with patterns for total population

The previous sections are useful for direct evaluation of density, but pertinent to evaluation of total population size only as correlative information. Consistent with the developing pattern for application of systemic management, the set of data needed to evaluate population size involves estimates of population size. When the management question involves sustainable total species-level population size, the consonant informative pattern involves measures of global total population size. This section evaluates human population size (not density, growth, location, or variation) using information on population size: simple numbers of individuals.

As usual, correlative information is involved, and it is probable that population size is related to body size as is density (Fig. 6.21; see also Appendix Fig. 2.1.22). Figure 6.24 shows the total human population (in numbers) to be much larger than the largest of all populations for other species of mammals with body sizes similar to that of humans (Appendix 6.4). The mean total population size for the nonhuman species shown in Figure 6.24 is 2.34 million, less than 0.04% of the current human population size. The current human population (about 6 billion) is about 2500 times more numerous than the mean of the nonhuman species of Figure 6.24. If the populations of these species are 10% of what they would be without the collective effects of abnormal human influence (i.e., including our population size, habitat destruction, global warming, pollution, energy use, and resource consumption), our species would still be 250 times more numerous than the mean in the absence of abnormal human influences. Based on data in $\log_{10}$ scales, the geometric mean population size for the other species is about 157,000. The human population is about 37,000 times this large (over four orders of magnitude larger).

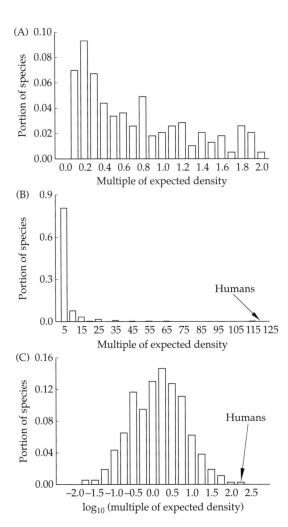

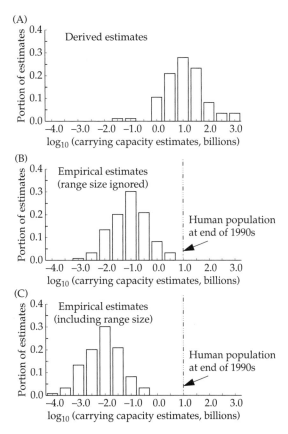

Figure 6.22 The frequency distribution of 368 mammalian herbivore species for density expressed as a multiple of that predicted from the regression line in Figure 6.21. (A) The subset that falls in the range of 0–2 for this multiple. (B) The position of humans in comparison to the frequency distribution for the entire sample, on a linear scale. (C) The position of humans in comparison to the frequency distribution for the entire sample on a $\log_{10}$ scale assuming humans were redistributed over 20% of the Earth's nonAntarctic land surface.

Figure 6.23 Three frequency distributions of estimates of the Earth's carrying capacity for humans (log billions). (A) "Derived" estimates from the combined data of Figure 6.20 (Cohen 1995b) and Table 6.1 ($N = 86$, mean of 6.75 billion) as estimates based on conventional approaches. (B) "Empirical" estimates, with a mean of 49 million, derived from 368 species of mammalian herbivores without considering human trophic level or geographic range size, but accounting for body size using the regression line of Figure 6.21 (Damuth 1987), and all suitable habitat available to humans. (C) The estimates in (B) shown shifted to the left to account for geographic range size (but not trophic level) and the 1996 estimate of the human population.

Figure 6.25 illustrates information similar to that of Figure 6.24 but based on biomass instead of numbers. In this case the mammalian species used for comparison are not restricted to those of human body size. The estimated mass for each species is biased because total biomass was determined by multiplying the population size by the mean adult body mass rather than the mean size based on prevailing age structure. Mean human body size was assumed to be 68 kg. Here, the mass of the human species is about 900 times larger than the mean mass among the nonhuman species in linear scale and 15,000 times larger in $\log_{10}$ scale.

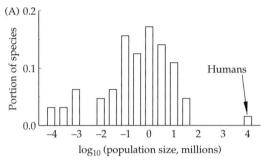

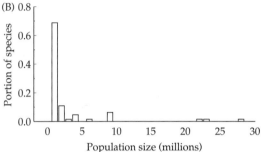

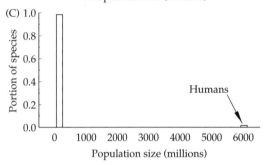

Figure 6.24 The human population compared to populations of 63 mammal species of similar body size (Fowler and Perez 1999; body mass and population approximations are from Nowak 1991 and Ridgway and Harrison 1981–1999). The data are shown in both $\log_{10}$ scales (A) and in linear scales (B, C). The range of population size spanned by nonhuman species (B) is concentrated into one bar when scales are expanded to include humans (C) with a bin size of 100 million (i.e., the single bar in C spans population size from 0 to 100 million).

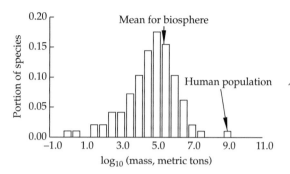

Figure 6.25 The mass of the human population compared to the mass of 96 mammal species represented by approximations for population size and body size in the published literature (Fowler and Perez 1999; body mass and population approximations are from Nowak 1991 and Ridgway and Harrison 1981–1999), and an overall mean for the biosphere.

Also shown in Figure 6.25 is the mean biomass per species if there are 45 million species and a total biomass of 1.1×10^{13} metric tons in the Earth's biosphere. The human population is about 1600 times larger than this mean. The challenge of dealing with overpopulation is emphasized by the national and international components of the problem (e.g., see Appendix 6.5 for an introduction to the variety of national-level challenges). Whether or not we should manage our population directly or indirectly will be treated briefly below.

6.1.7 Other species-level characteristics

Several more examples of relevant species level characteristics are presented below to further illustrate how information on the limits to natural variation can be used for evaluating the human species, setting normative standards and establishing management goals for human action. Such examples help emphasize the:

• Number of dimensions over which comparisons between humans and nonhuman species can be made
• Synergistic/correlative effects of interrelated characteristics
• Complexity of implications for management
• Importance of human overpopulation
• Degree of risk our species faces, and poses for other biotic systems

6.1.7.1 Density dependence and rate of increase
Having temporarily broken from the natural constraints of population limitation normally imposed by ecosystems (Catton 1980) and the biosphere (and

If we insist on an abnormal consumption of energy at the individual level (to maintain what we think of as acceptable standards of living—an opinion on our part that plays into decision-making and what we see; Figs 1.1 and 1.4), it can best be attained with a reduction in the population to follow the diagonal line through point f to the left and upward (smaller population, increased per capita consumption). This would be effective systemic mitigation rather than placing the burden of human abnormality on other species, ecosystems, or the biosphere—taking advantage of tradeoffs within our species. Note the challenge we encounter here, however: increasing our standard of living (e.g., plentiful food, material wealth, and accessible/quality medical services with the energy they require) has a strong tendency to result in the mechanical reaction of increased (rather than decreased) population level. This exemplifies the dilemma (lack of consistency) faced by philanthropic organizations that simultaneously provide funding for dealing with population issues on one hand and quality-of-life issues on the other.

In general, bounded spaces such as that shown in Figure 6.31 are merely crude representations of those observed in nature (recall Fig. 2.34) and occur within systems where such bounds involve many dimensions. Research to define such a space is an extremely important exercise in finding the limits to a set of sustainable options and the abnormality in extremes where we can identify realistic action (management of the human). The difficulty experienced in contemplating such action can be translated to empathy for other species in their being forced to change in reaction to our abnormality—understanding not often brought to conventional management realistically.

In the case represented by Figure 6.31, it is clear that populations can be abnormally small— something that might result from systemic reactions to current human abnormality (e.g., an extreme pandemic), but not something we would deliberately seek any more than we would restrict per capita consumption to unrealistic levels. Clearly, a human population of less than 4000 would never be thought of as sustainable (we would be on our own endangered species lists) and a standard of living confined to simply meeting metabolic needs is likely to be rejected by most (an opinion and real force behind our abnormalities). As in all cases, sustainability is not to be confined to a single value for either population or per capita energy consumption. Data on the variance of per capita energy consumption for other species would be useful to establish the limits to normal variation in this measure of species. A human population of 9.62 million (which maximizes biodiversity for population size under current circumstances, Fowler 2008) is intermediate to the extremes represented on one end by today's population and on the other end by the lower 95% confidence limit among nonhuman populations. Maximum sustainability is likely to be found close to where lines g, h, and e cross in close proximity. This is where diversity is maximized for both total energy consumption and population size. It involves a normal per capita energy consumption. Other measures of diversity might result in even better consistency— something actually achieved in nature, of course. A population reduced to the mode among nonhuman species otherwise similar to humans (Fig. 6.24 and bottom panels of Fig. 6.31) would allow for a bit of freedom to maintain a standard of living (involving per capita energy consumption) close to that of the current global average but much less than enjoyed, and taken for granted, in countries such as the United States. Simply because we enjoy something does not make it sustainable.

The following sections and Appendix 6.1 give further examples of the factors involved in the complexity of interconnections, including management with applications at the individual, species, and ecosystem levels. They provide examples of information we often attempt to use in conventional management involving connections such as those made explicit in Figure 6.31. Knowing about individual connections or relationships involves understanding as steps in the right direction but on an endless and impossible path.[14] Although they provide motivation for action, they are not the basis for sound goals. Although they relate directly to human population issues, we have yet to decide what specifically to do: reduce the human population to solve other problems identified in this chapter (Plate 6.2), solve those problems knowing that the human population would respond, or let nature take its course. Before addressing this issue, the following sections are intended to help point out a very few of the kinds of things that might be involved. Decisions must be made by society knowing that there is a composite of risks and benefits involved in the complexity (the infinite of Fig. 1.4) behind empirical information in species-level patterns. Known connections with overpopulation include the risk of catastrophic changes such as starvation or disease spreading to be even more extensive and intensive than it is today (malnutrition involving over 50% of our population, Pimentel *et al.* 2007). Alternatively, or in conjunction with such events, we may find ourselves in the population-reducing consequences of war over ever-increasing scarcity of renewable resources such as food and water, especially following further concentrations of humans fleeing the effects of global warming. The connections often involve forces or factors that contribute to population reduction and concomitantly themselves respond to population reduction—a reciprocity exemplified by the connection between CO_2 production and population. Of particular concern is the current anthropogenic extinction event (Appendix 6.6), reminding us of the risk of human extinction as an unintended consequence of what we are doing now and as a result of what we have evolved to be including our belief systems and all the other aspects of being human.

6.2.1 Individual level

We are only beginning to understand the problems caused by overpopulation, and the corresponding benefits of a smaller population, measured either in terms of abnormality or conventional value systems. Millions of people have direct personal experience of poverty, starvation, disease, and conflict over dwindling resources (including water, food, fuel, and land). Millions experience other indirect results—increased competition for jobs and housing due to immigration, increased crowding, conflicts over territory and resources. To the extent that these problems would decline with reduced population, each individual would experience fewer (and, at the population level, a smaller portion of individuals would experience them).

Any action(s) that contribute to reducing human population to sustainable levels could take centuries,[15] or could occur suddenly due to one or more catastrophic events, which are all too easy to imagine and with increasing chances of happening. In either case (barring things like widespread radioactive impacts), a smaller population would relieve surviving individuals from a variety of stresses caused by daily life on an overcrowded planet. Food, water, shelter, and space would be abundant, or adequately available; but only on average (there will always be issues regarding social inequity—variance in such things is also impossible to avoid, but abnormality in such matters is a systemic problem, well beyond the focus of this book). Vulnerability to pollution-related diseases such as cancer, heart disease, and diseases related to compromised immune systems would decrease owing to reduced or eliminated use of chemicals. There would be a reduction in the transmission of communicable diseases through human contact. With a reduction in the widespread monoculture of human overpopulation, we might expect a reduced rate of emergence of new diseases through evolving virulent strains. Individuals would, on average, benefit from more spiritual, emotional, and aesthetic experiences of a more normal natural world and more contact hours per day with that world. The psychological benefits are unfathomable.

A controlled approach to reducing population to sustainable levels would require individual sacrifices for the sake of future generations of all forms of life. A question now before us is: Are those sacrifices to be made in direct control of our individual survival and mortality (things we consider at the roots of basic human rights) or are they to be made in reducing or eliminating the abnormal uses and consumption of energy, chemicals, and resources in a way that systemic change results in reduced human population? A third alternative is an uncontrolled population reduction brought about systemically (starvation, disease, wars, or various combinations of such factors) in reaction to conditions brought about by current and past management. This is where different patterns enter as insight, if not for guidance. Very few if any other species willfully sacrifice their survival or reproduction so altruistically as to be for the benefit of the system of which they are a part. The pattern here is one in which limits are established systemically—the whole (ecosystems, biosphere) brings controlling forces to bear on the parts (species, individuals). These patterns are basis for arguing that changes in our survival and reproduction need to be the indirect result of our matching the patterns of other species in energy consumption, chemical use, geographic range size, and resource consumption (all the ways we are abnormal in our interactions with other species and ecosystems) to allow systemic reduction of our survival and reproductive rates. We are confronted with the depth/magnitude of our challenges and the realization that it is highly unlikely that we will willfully achieve sustainability. Although the game of sustainability may be an infinite game (Carse 1986), and one we could play (i.e., there is hope), the strength of socioeconomic forces, our genetic nature (Table 3.1), our ways of thinking, and other aspects of what we are as a species dictate that science will document the unfolding of our reality (even our own extinction) more than provide information that gets used for achieving real sustainability. Birth control (a sacrifice of having fewer children) is not a "fit" option, not only because it is inconsistent with patterns among other species, but because of the (at least potential) genetic effects of exerting "intelligence" in making the decision to limit family size to reduce the population (individuals making decisions about fitness is all too reminiscent of Hitler's actions). The same holds for mortality—decreased longevity is not an option among individuals for self control at the species level.[16]

Other sacrifices individuals are likely to experience with reduced population levels (still involving human values) might include reduced use or availability of technology (either as a result of reduced population or as a contribution to its reduction). Culture would change; some forms of cultural expression would be sacrificed and others enhanced.

In achieving change, choices by individuals will be paramount. There are so many political, religious, social, racial, educational, economic, and psychological issues to deal with that the task of human-managed social change (as opposed to change systemically imposed by nature) seems

nearly impossible. It may be impossible (as, in so many cases, bringing us back to combinations of Table 3.1 wherein evolved characteristics lead to extinction). Some of the complexity of reality is impressed upon us when we consider the meetings, brainstorming, consideration of options, and decisions for action that will have to ensue worldwide when and if we remember collectively that despite our apparent supremacy, our species is only a tiny part of the universe within which we are trying to find a sustainable existence.

The likely benefits to the individuals of other species in a reduced human population are also real. These realities are impossible to overemphasize (although very possible to overvalue as is done in biocentric views of what we need to do). However, it is beyond the scope of consideration here. Nevertheless, these benefits (primarily reduced abnormality from the perspective of systemic management) must be acknowledged as counting among changes that would be achieved in restoring humans to normalcy. Taking systemic action is always a matter of extending our compassion, and concepts of justice and fairness, to all individuals of all species. Equity in the form of neutral fitness may prevail at the core of natural patterns.

6.2.2 Species level

Risks to the human species (i.e., as a species, not as individuals) in reducing population density are minimal compared to the benefits—both inherent to reduced abnormality. Some risks of overpopulation to humans at the population or species level have been temporarily averted or delayed in conventional management—but at the expense of other systems (e.g., degraded ecosystems and ultimate associated human risks). The elevated human population has been made possible, in part, by reliance on fossil fuels as a source of energy outside the living ecosystem (Catton 1980). To the extent the depletion of fossil fuels can be postponed, they can be used to help support more than a minimal standard of living while the population is reduced, a standard that could hopefully be maintained afterwards,[17] but for which there is no guarantee. Long-term benefits of population reduction at the species-level include diminished risks of extinction from diseases, loss of food, or loss of other goods and services provided by ecosystems but lost in their destruction or disturbance. Exacerbating the risk of extinction, in this regard, would be factors intrinsic to our species such as wars or other conflict stimulated by reactions to ecosystem degradation. The rates of evolutionary reactions among other species to human presence, such as pesticide resistance and emergent diseases, would likely decline, but details are impossible to predict.

While there is some risk of extinction among species that currently require human support (e.g., domestic species and pets), the majority of effects experienced by other species would be a reduction in human-related factors that currently contribute to elevated risk of their extinction. Human overpopulation is one of the contributing factors behind the numbers and extinction rate among endangered species (Appendix 6.6). Overpopulation clearly contributes to the overutilization of resources with its connections and contributions to risks (abnormality) for other species.

6.2.3 Ecosystem level

At the ecosystem level, the experience of human abnormality is probably not a terminal problem. Following any reduction of the human population or any complete removal of humans, ecosystems will experience the opportunity to rebuild from the remaining species as they have from numerous mass extinctions of the past (Rosenzweig 1995). For ecosystems, a reduction or elimination of the human population is equivalent to curing an ecosystem-level disease (Hern 1993).[18] Assuming that ecosystems will survive this event, they will likely be more immune to similar events in the future—part of the development of an ecosystem-level phenotype with a species composition more resistant to the overpopulation of any component species than is now the case. Over time, similar (albeit, less extreme) disruptions[19] have been survived by ecosystems and, in the process they may have become better at dealing with such problems (e.g., through the selective prevalence of diseases and parasites, as an ecosystem-level parallel to leucocytes or immunity).[20] In comparison to natural rates of change, the rates of loss of diversity, and other ecosystem-level changes (MEA 2005a,b), are parts of the ecosystem experience of human overpopulation and all of the abnormality it brings with it in terms of our interactions with ecosystems. Each would be alleviated by a smaller human population and progress would be made toward the objective of ecosystems with normal attributes.

6.3 The eastern Bering Sea example

The commercial fisheries of the eastern Bering Sea produce food to help meet human needs. Such benefits are recognized as part of what are referred to as ecosystem services. As with all ecosystems, the eastern Bering Sea provides ecosystem services to all of its species. Of the overall production within this ecosystem, each species consumes a part. Very importantly, there is a part of this production that each species does not consume. The part that is not consumed is available for the other species. These dynamics involve a web of trophic interactions; trophic interactions characterize ecosystems with every species of that ecosystem involved. Similarly, each species is involved in coevolutionary interactions with other species so that there is also a web of genetic interactions. Both the trophic and evolutionary webs involve higher-order interactions and feedback—coevolution occurs in indirect interactions as well as in direct interactions. These webs cross hierarchical scale.

In Chapter 1 and above, patterns were introduced representing the amount of biomass consumed within the eastern Bering Sea ecosystem by the consuming species. Frequency distributions for this measure of species have been presented for consumption from individual species (Figs 6.1–3), groups of species (Fig. 6.8A), and the entire eastern Bering Sea ecosystem (top panel of Fig. 6.10). This information can guide decisions about what level of biomass to harvest in commercial fisheries as well as our allocation of harvest among alternative species (Fig. 2.8, Fowler 1999a). Similar data can be used for allocating harvests over space (Fig. 2.13) and setting aside marine reserves (Fig. 2.15), or making seasonal allocations (see Fowler and Crawford 2004 for a partially consonant example, based on data for consumption by seabirds as presented by Crawford *et al.* 1991 illustrating the process if data for mammals were available). Using data that show the limits to natural variation in patterns consonant to questions regarding sustainable consumption from the ecosystem (e.g., Figs 6.8, 6.10, and 6.11 for biomass consumption) would be an example of applying systemic management directly at the ecosystem level, i.e., addressing the ecosystem-level question of what levels of biomass extraction are sustainable.

Regulating our consumption is only one aspect of management; not consuming is another part—what is not consumed is left for the remaining species of the ecosystem, for the ecosystem as a whole, and for the biosphere. At the individual species level, part of the biomass produced by a resource species is not consumed by each consumer (e.g., production by walleye pollock that is not consumed; Hobbs and Fowler 2008). Future research can determine the portions of production that are not consumed by each consumer species when the production is that of multispecies groups or full ecosystems. Such data would be consonant with management questions regarding production that should not be consumed whether it is production from any individual species, species group, or entire ecosystem. Such information serves to guide us in insuring the sustainability of the system for all species—full scale sustainability. As usual, management based on such information is consistent with the complementary information regarding what can sustainably be consumed (Hobbs and Fowler 2008).

6.3.1 Anthropogenic influence

As with all patterns, the patterns we observe in the eastern Bering Sea today reflect both current and historical abnormal human influence. These patterns, and the trends within them, are, in part, the product of all anthropogenic activities. Systemic management has the objective of making those circumstances free of human abnormality. In the meantime, existing patterns provide one means of accounting for anthropogenic influence: it is inherent to the information they embody (Fig. 1.4).

Anthropogenic factors also influence the quality of information represented by patterns in another way. Human limitations not only prevent accurate

projection of future circumstances, they also influence our observation of patterns. All information is subject to errors of estimation and other statistical imprecision—the anthropogenic factors of imperfect science. The information in hand is a start, but it is not perfect.[21] Part of attaining useful scientific information is the matter of science conducted with as much quality as is possible.

In the eastern Bering Sea, the patterns useful in managing the walleye pollock fishery above (Figs 6.1–3; see also Fowler 2008, Hobbs and Fowler 2008) represent results of the kind of science needed in management. These patterns reflect abnormal human influence realized through past fishing, whaling, toxic wastes, global warming, marine acidification, and sealing. The matter of anthropogenic influence is involved in these patterns, whether it involves patterns regarding predation rates on individual species, finfish as a group of species, or for total biomass consumed in the ecosystem as a whole. Integrative patterns and quality science are two aspects of accounting for anthropogenic influence.

A third component of accounting for anthropogenic influence involves correlative relationships. Comparing patterns across space, time, and systems makes it possible for scientists to directly account for a variety of factors. These include the variety of anthropogenic factors. Thus, as patterns vary in correlation with harvest rates, pollution levels and habitat modification can be used to directly incorporate anthropogenic factors—always in regard to patterns consonant with the management question so that the correlative relationships are pertinent to refined questions. As in other ecosystems, a great deal of research will be required in the eastern Bering Sea to find and use such patterns.

A fourth part of accounting for human influence directly involves the element of adaptive management introduced previously. We could, for example, be precautionary and reduce our harvest of fish biomass to levels even less than indicated by existing species-level patterns for consumption while obtaining more accurate and complete data. Better determination of the statistical central tendencies of species-level patterns, or measures of maximized biodiversity, might result in a decision to harvest more or less than indicated initially. There are other factors that influence observed patterns, however, and, to be systemic in approach, we would also reduce our use of polluting chemicals, our production of CO_2, and consumption of energy—involving all influential factors directly or indirectly involved. In other words, we would endeavor to fit into our world normally in every conceivable way not simply for the benefit of the eastern Bering Sea but for all ecosystems. During such management (and following, if we have survived systemic forces that work toward the same ends), we would monitor to observe changes in the frequency distribution for biomass consumption among nonhuman species to see if the patterns shift to show higher or lower advisable harvest rates, recognizing that this can involve many decades. This monitoring would also help document interannual (short-term) variation and related relocation of species within patterns. Similarly, over longer time scales, patterns in relation to regime shifts in the climate, and across seasons will add specificity (and allow for addressing more refined management questions).

6.3.2 Other correlative factors

The complexity of ecosystems and environmental conditions requires considering other elements of importance. Species-level features are often correlated (Chapter 2) and, in our management oriented research, we need to choose species otherwise similar to humans as a part of taking ourselves into account (Management Tenet 1). Body size is one element to account for, as illustrated above in relationship to consumption rates, geographic range, CO_2 production, and population numbers. Trophic level (both human and resource) is to be taken into account as was done in consideration of population numbers and geographic range. Environmental conditions such as climate are to be taken into account (i.e., we know that empirical patterns in harvest rates and allocation across resource species changes with such conditions; Melin *et al.* 2008). Other questions relate to different dimensions:

• How many resource species should be harvested?
• What should be the seasonal and spatial allocation of harvests?

• What portion of the ecosystem should be set aside in marine protected areas?

Answers to these questions all can be represented in patterns among species, most of which have yet to be described for the eastern Bering Sea ecosystem, and many of which will involve correlative sub-patterns.

As previous examples have shown, there are two ways forward and it is advisable to proceed on both fronts simultaneously. One, of course, involves research to develop, better describe, document, and understand the relevant species-level patterns for any ecosystem such as the eastern Bering Sea. This involves conducting the kind of biological/ecological studies that produce information to reveal patterns such as those seen above and in Chapter 2, specifically comparing humans to other species as done in this chapter. The depiction of species-level patterns (see methodology in Appendix 1.3, Fowler and Perez 1999) can be developed for metrics such as the number of species consumed as they are related to other factors such as body size, trophic level, or metabolic rate. Adaptive approaches would most likely be necessary to account for past human influence on all fronts, but reconstructive modeling and estimates of populations of marine mammals prior to human intervention would be helpful (useful for gaining insight, but, in management, important only if used for producing information consonant with management questions).

The second approach complements the first and involves inter-ecosystem comparisons and ecosystem comparison over time. The same kinds of information on the limits to natural variation as listed above would be found for other systems. For example, the numbers of species consumed is a feature of ecosystems studied intensely in food web research and many relevant data and patterns are already published. One advantage of this approach is that, for a few ecosystems, data have been archived in existing publications and data bases. Disadvantages of both approaches include the logistical difficulty of collecting new data; monitoring is critical to any form of management (Management Tenet 8).

Comparisons across ecosystems are exemplified in Figure 6.32. This figure shows a pattern in which the normal aspects of size selectivity are correlated with body size. Not only is this a biosphere-level pattern to engender and support ecosystem-based management (ecosystems being part of the biosphere), it is also an example of information that (like that of Fig. 6.29) also opens the door to direct application of evolutionarily enlightened management. Patterns regarding things like harvest rates account for evolutionary dynamics (being integral in nature). The pattern depicted in Figure 6.29 exemplifies information that allows for directly addressing questions of selectivity. It again points to the need to directly involve as much as possible in correlative information (Fig. 6.32 involves body size again). An ecosystem-based approach would result in a regression line for commercial fisheries in harvests of all species (heavy dashed line) that corresponds to that for marine mammals (solid line). A single-species approach for walleye pollock would make use of data like that in Figure 6.33 (similar to that for cod in Fig. 6.29, part of the overall collection of data for a variety of species presented in Etnier and Fowler 2005).

Both approaches (characterizing specific ecosystems, and making ecosystem comparisons) would be possible while waiting for systems to respond to relief from abnormal human influence (e.g., reduced commercial harvests, reduced CO_2 production, reduced pesticide production), and while conducting research to provide more accurate characterization of consonant patterns. As in all cases, it is important that much larger samples be made available than we currently have in hand.

6.3.3 Control rules in fisheries

Conventional management of commercial fisheries involves decisions and policies produced by committees, panels, managers, and stakeholders—often in a political context (top row of Fig. 1.1). Several of these are embodied in what are sometimes called *global control rules* (e.g., McBeath 2004, Restrepo *et al.* 1998). The essence of these rules as conventionally applied are depicted in the top panel of Figure 6.34. Involved are the rate of harvest at "virgin" population levels of the resource species (the maximum), the degree to which a resource population is reduced (along the abscissa), and the reduction in

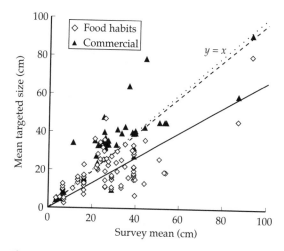

Figure 6.32 The mean size of individual prey items taken in 85 cases of commercial harvesting and marine mammal food habits as related to the mean size of prey/resource populations (from surveys of commercially valuable fish populations). Solid line is linear regression for food habits means, heavy dashed line is linear regression for commercial means, and light dashed line represents $y = x$. The slope of the regression line relating dietary data to size is statistically significantly less than 1.0 ($p < 0.001$, Etnier and Fowler 2005).

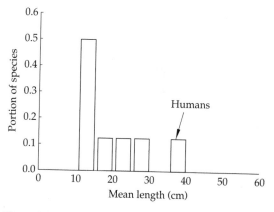

Figure 6.33 The mean size of walleye pollock taken in commercial fisheries (n = 21) compared to the mean size of walleye pollock taken by 7 species of marine mammals (from Etnier and Fowler 2005). These data represent a subset of the data shown in Figure 6.32.

harvest rate mandated to protect the resource from overharvesting (the ordinate). Population reduction is carried out, in conventional approaches, taking into account our knowledge that doing so usually stimulates production. Reductions to 40% of unharvested levels is a common "rule of thumb" based on patterns in density-dependent relationships in the population dynamics of species like those being harvested. Population reduction is not accompanied by a reduction in fishing rate until a threshold is reached. In other words, in fisheries management, current control rules permit, or establish by policy, a value of 1.0 for the length of line A in the top panel. Initial harvest rates (fishing mortality on "virgin" populations) are established on understanding of population dynamics

(particularly the concept of MSY; Restrepo *et al.* 1998) which, as seen in the discussion of Figure 5.6, often involves total mortality rates (M) estimated for the resource species. Target population levels are also established on the basis of our understanding of population dynamics, specifically the population level that would result in the maximum rate of change in recovery from population reductions. The protective reduction in harvest at low resource population levels is rather arbitrary but attempts are made to reduce harvests to zero for populations at levels between 0.0 and 20.0% of "virgin" levels (K; the solid lines in Fig. 6.34 would drop to 0.0 at 0.2 on the abscissa).

Systemic management would deal with each of the components of control rules individually with consonant empirical patterns (Fowler *et. al.*, in preparation). We have already seen the kinds of information that would be used to establish the upper limit to harvest rates (e.g., Fig. 5.6). Such information leads to guidance that reduces the upper limit to small fractions of harvest rates used in conventional management (often to less than 10% of conventionally established rates—7.0%, 3.3%, and 5.9% of recent harvests for hake, herring, and mackerel, respectively, as seen above; 16.8% if we use the regression line of Fig. 5.6). Systemically established rates are exemplified by the upper asymptote of the curve in the lower panel of Figure 6.34 assuming 10% of conventional rates applies (actual values would be based on information such as that in Fig. 5.6—5.4% if we used the mean of values for the species just mentioned).

In systemic management, the degree to which a population is reduced is not guided by the intention of stimulating production (line A in Fig. 6.34), even though such a reaction is probably often one of the results (this is a pattern observed and characterized in past research). Instead, observed population reductions would serve as the consonant guidance; that is, the intent would be to mimic natural patterns in population reduction rather than to stimulate production. As Pimm (1991) has shown, nonhuman predators often have effects in which prey populations actually increase (instead of decline); there is a great deal of variation in population changes as a result of the effects of predation (Fig. 6.35). Thus, in the full suite of applications of

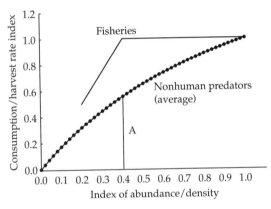

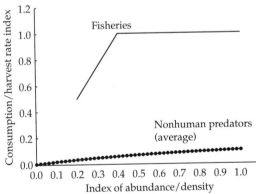

Figure 6.34 A comparison between functional response curves and control rules in fishery management (Fowler *et al.* in preparation). The top panel compares the shape (the maxima are equivalent), and the bottom panel compares both the shape and the magnitude of allowable harvests (hypothetically chosen to be 10% of those in conventional management, more than all of the values most sustainable for hake, herring, and mackerel presented in the discussion of Fig. 6.7). See text for details regarding the shape of the response curve (length of line A).

systemic management, there would be an average reduction of about 10% (i.e, the populations would be reduced on average to about 90% of their "virgin" levels, rather than more consistently to 40.0% as practiced in conventional management: Fowler *et al.*, in preparation). Here, we are back to the patterns in variance and limits to variation in variation (the standard deviations of a variable are not fixed quantities and vary within limits themselves). Thus, instead of an intentional reduction to within narrow limits, management would reduce populations to varying extents depending on circumstances (i.e., a single value of 60% reduction to 40% of "virgin" levels would be avoided, even as a "rule of thumb," owing to the fact that such a policy would result in variance which would be much less than that shown in Fig. 6.35). Future research would examine the correlative components of the variation shown in Figure 6.35 to better identify advisable reductions for species with specific characteristics (e.g., features of their life-history strategy, body size, and trophic level—a wealth of opportunity for research). Such research would also involve environmental factors such as climate, latitude, and depth. It would include explicit study of anthropogenic influence.

Finally, the shape of the curve used to enforce restrictions on harvest rates as a function of

population size would be based on information such as that shown in Figure 6.36 (Fowler *et al.*, in preparation). Here the average rate at which nonhuman consumers consume resources when resource populations are at 40% of their normal or "carrying capacity" levels is about 56% of the corresponding maximum. The general shape of the resulting "generic" curve is shown in the bottom panel of Figure 6.34. As is the case for population reduction, the pattern in Figure 6.36 is informative regarding what we would strive for on a global (biosphere or even ecosystem) level, but open to a great deal of research regarding correlative factors to enable application in local situations, and for specific species.

As in all cases, information we have in hand is information collected from systems subjected to abnormal species-level influence by humans (all the ways we are demonstrated to be abnormal, as shown in this chapter, plus those ways for which we have not made measurements and comparisons). Better information will come from studies of systems relieved of human abnormality

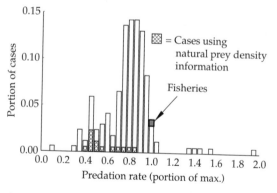

Figure 6.36 The pattern observed empirically in the length of line A in Figure 6.34 ($N = 170$). This measure is the ratio of predation rates measured when prey populations are at 40% of maximum levels to that observed at virgin populations' levels (population index of 1.0 in Fig. 6.34). The value used for line A in Figure 6.34 is the mean for consumption rates among 13 cases in populations reduced to 0.4 of the population levels observed under more normal circumstances (population levels for resource species at which fisheries are often managed to have a fishing mortality rate equivalent to that implemented for "virgin" populations, or $A = 1.0$—the latter often close to the total natural mortality rate, M, for resource species as described for Fig. 5.6).

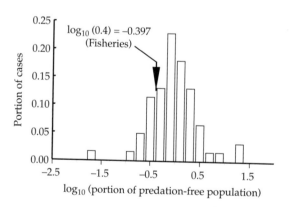

Figure 6.35 The pattern in predatory effects on prey populations shown as $\log_{10}$ of the ratio of the altered population density to the original population density (e.g., 0.5 corresponds to a population reduced to half its original density, $\log_{10} 0.5 = -0.301$). This sample involves 61 cases of predator/prey pairs across a wide variety of taxa, trophic levels, and locations (from Fowler *et al.* in prep.).

whether that happens as a result of management or through systemic reorganization (which would include the self-organizing dynamics of complex systems [Ahl and Allen 1996, Brown 1995, Burns *et al.* 1991, Callicott 1992, 2006, Kauffman 1993, Levin 1999, Lewin 1992, Maturana and Varela 1980, Norton 1991, Rapport 1992, Rosen 1987, Waldrop 1992], but would include the contextual factors of inclusive hierarchical systems) to force humans into more normal roles within ecosystems and the biosphere. As in other cases, however, current data are indicative of sustainability that accounts for the anthropogenic influence expressed.

6.3.4 Marine protected areas

What portion of the eastern Bering Sea should be set aside in marine protected area (MPA) status? Figure 2.15 shows the pattern consonant with this question. As pointed out by Fowler and Johnson (in prep.), the matter of excluding humans from any particular region of this ecosystem is probably not as important as reducing harvest rates, pollution, or CO_2 production. This conclusion is based on the fact that almost any portion we might choose for MPAs falls within the range of variation exhibited by other species in the portions of the ecosystem outside their geographic ranges. We could choose to use the mean of areas not occupied by other species (36%) and decide to set aside 533,000 km^2, but other options are not unacceptable.

The northern fur seal population is declining and classified as depleted under terms of the Marine Mammal Protections Act (Towell *et al.* 2006). What portion of this species' geographic range (within the eastern Bering Sea) should be set aside in MPAs? The northern fur seal occupies the full ecosystem to one extent or another; this leads to the same conclusion as reached above for the full ecosystem.

What portion of the geographic range of bearded seal (*Erignathus barbatus*) should be set aside in MPAs? The map shown in Plate 6.3 shows the geographic range of the bearded seal in its overlap with the eastern Bering Sea. Also shown are subsections of the bearded seal's range within the eastern Bering Sea as defined by varying numbers of other marine mammal species ($N = 20$) with geographic ranges overlapping that of the bearded seal. Thus, on a species-by-species basis, the portion

of the bearded seal's range that is occupied (where there are overlapping geographic ranges) can be calculated. Similarly, the portion of the bearded seal's range within the eastern Bering Sea that is not occupied can be determined (i.e., if 40% is occupied or occurs in the overlap of the geographic range of a particular species with that of the bearded seal, then 60% is not occupied). Figure 6.37 shows the frequency distribution for the portion of the geographic range of the bearded seal within the eastern Bering Sea not unoccupied by the other marine mammals in this ecosystem (mean = 0.47 or about 346,000 sq km). Again, there is a broad span of options without being abnormal (it is essentially impossible to be abnormal). Nevertheless, we could choose to set aside marine protected areas with a total equivalent to the mean of unoccupied area among the nonhuman mammalian species. Or, in this case, picking the mode (50%) might be an advisable option. This exercise can be repeated for every species in this ecosystem (and, of course, for every ecosystem).

Overall, there is no basis for concluding that there is abnormality in the area we have designated for protection within the eastern Bering Sea. Harvesting spread across the entire ecosystem cannot be seen as abnormal (Fowler and Johnson in prep.). In contrast, the distribution of biomass consumption and the level of biomass consumption are distinct issues for which abnormality may be a problem (e.g., Figs 6.1, 6.8). Thus, the real problems (i.e., human abnormality) are found in the rates at which we are harvesting, in the selectivity of our harvests, in the production of polluting chemical compounds, in producing CO_2, and in other ways that have yet to be identified, and measured and illustrated, as above in this chapter and elsewhere (e.g., Fowler 2003, 2005, 2008, Etnier and Fowler 2005, Fowler and Hobbs 2002, 2003, Hobbs and Fowler 2008).

Part of what managers are trying to deal with when addressing questions of protected areas is the distribution (allocation) of harvests over space. Fowler and Crawford (2004) demonstrate how this would be done, using data for birds, when data for mammals would be more appropriate (the data for birds would be very helpful in looking at correlative relationships across various species-level characteristics). For management of the distribution of fishery

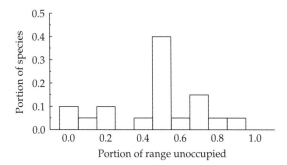

Figure 6.37 The pattern in the portion of the geographic range of the bearded seal in the eastern Bering Sea that is not occupied by each of the other marine mammal species ($N = 20$) based on information shown in Plate 6.3.

catches within the eastern Bering Sea, the density of individuals for species represented in the map of Plate 6.4 would be used to calculate their consumption as spatially distributed within this ecosystem and used to allocate fishery catches. Changes in response to (correlated with) changes in climate would be used to account for such factors directly.

6.3.5 Asking new questions

While we await information from systems free of abnormal human influence, systemic management would extend management questions to other realms and ask new management questions:

• Should we transport harvested food from the eastern Bering Sea to other ecosystems?
• How many ecosystems should we be using for resources (living space, or solid waste disposal)?
• How much nitrogen (or any other element) should we be using (or releasing to the environment)?
• What is a sustainable geographic range for our species? Should it include the eastern Bering Sea?

Complexity involves many such questions; finding consonance between these questions and the patterns that inform decision making is crucial.

6.4 Summary and preview

This and the preceding chapters have 1) laid out the basis for, 2) illustrated the choices of scientific

information to be used in, and 3) provided examples of, systemic management. Systemic management was shown to be an approach that can be applied not only to ecosystems, but also individual species, groups of species, communities, and the biosphere. We could call the ecosystem application "ecosystem management" but the unrealistic connotation of control over ecosystems argues against such terminology (Management Tenet 2), as does the breadth of application across the various levels of biological organization (Management Tenets 3, 4). We could call it "ecosystem-based management" but this implies loss of "individual-" and "species-" based management—all of which have to be part of management to be realistic. Calling it "systemic management" or "reality-based management" implies transcending hierarchical boundaries between the various levels of biological organization, including individuals, and applying to them all simultaneously. Systemic management significantly reduces conflicts prevalent in conventional forms of management. This is because the elements of conflict are dealt with inherently in the kinds of information used. Such information is carefully chosen from the spectrum of research results produced by the biological sciences—information that is consonant with management questions. It requires no conversion, thus obviating conflict. What was basis for conflict (top row of Fig. 1.1) becomes basis for asking clear management questions (bottom row of Fig. 1.1).

Chapter 4 presented a critique of conventional management to clarify its inadequacies for comparison with the advantages of systemic management. Chapter 5 put forward an evaluation of how well systemic management meets the needs for management as identified in the literature and suggests how it can be implemented when guided by information on the limits to natural variation when the information is consonant with the management question being addressed. This chapter used specific examples to add clarity to the issue of consonance between management question and guiding patterns to more clearly define the kinds of biological/ecological research needed to provide guiding information. In the epilogue, we visit the issues of management a bit more philosophically than in the previous chapters.

CHAPTER 7

Epilogue

Without hope, all we can do is eat and drink the last of our resources as we watch our planet slowly die. Let us have faith in ourselves, in our intellect, in our staunch spirit.

—Jane Goodall

Confined to science, this book might have done little more than reveal some of the extreme abnormalities of our species in cold statistical terms (e.g., the graphs of Chapter 6). It might have stopped with attempts to explain patterns among species—especially the pattern of human abnormality. The underlying message might have been a prediction about how things are going to be after the homeostatic forces of nature return sustainability to ecosystems and the biosphere—with humans either extinct or in more normal positions within interspecific variation. Confined to science, this book would have been part of the growing awareness of ourselves as a species, comparing ourselves with other species as part of our unfolding story—maybe to be marked by a species-level rite of passage.

As easy as it is to predict that we will suffer for our hubris and arrogance, however, it also seems that we hold the potential of taking self awareness as a species to the level we occasionally do as individuals. We are likely the first species with any self awareness regarding the systemic consequences of our actions—impacts that make us a malignancy in the biosphere. Self-restraint may be an option for us as a species; on many occasions we achieve self-restraint as individuals, even in the face of impulses, urges, and emotions that often get us into trouble. Blocking progress, however, is an artificial wall between reality and management that is unnecessary, and counterproductive; current passage through this wall results in a needless gap between science and action at great cost to all forms of life. This impediment is a product of our thinking, beliefs, habits, and their evolutionary origins. These things are obviously

not alone, but they count among the factors contributing to our abnormality as a species and the consequences for others. However, thinking, beliefs, and habits are also things wherein we have hope of making change—taking responsibility as a species. I would have found it unfulfilling to stop with nothing more than a presentation of the science, even irresponsible. Certainly, confining myself to science would not have met my official responsibilities! Realistic management is something we aspire to achieve whether as individuals or as a species. Knowing that the two can be diametrically opposed when conceived in simple terms is a major step forward. Knowing that the paradox is not insurmountable is an even more encompassing step.

Our species' predicament in today's world can be characterized as an addiction. We exhibit dependencies on energy and resources to the detriment of other species and ecosystems that also occupy this planet. Not only have species gone extinct at the hand of man, but full ecosystems have fallen victim to our dependencies. It is well known that addicts, as individuals, have a limited chance of recovery, even with the support of fellow addicts and professional counsel. Can our species, at the species level, undertake a successful self-intervention? This book is my attempt to provide information that looks to other species for help, as mute and unconditional as they are in their interactions with us.

A wide variety of experts and organizations have worked very hard at defining management so as to expand beyond what we do as individuals, and to progress beyond what we do in our interactions with other species. Experience has shown that the results of these efforts are essential; in many cases, the resulting advice has been around for centuries as part of ancient wisdom traditions (Hobbs and Fowler 2008). Across time, a variety of tenets have emerged. When we accept these tenets collectively, the words of Wordsworth apply: "Come forth into the light of things, let nature be your teacher". Progress toward objectivity cannot be achieved, however, if we do not reject one tenet repeatedly found in the history of our thinking: putting stakeholders in the position of attempting the impossible—converting one kind of information to another, and, in the process, making values other than that of sustainability the basis for policy. We have to face, accept, and work with reality; the laws of nature cannot be violated—the patterns of nature, as products of reality, provide guidance. There is no suitable alternative that I know of—nothing that offers comparable objectivity. Can we achieve sustainability or are we going to suffer as victims of being, doing, and thinking that are incompatible with reality?

Like anyone who seeks to make a difference and to make the world a better place for all forms of life, I encounter the fact that change in the human realm depends, in part, upon what individuals can accomplish. Change, as a species, is critical and always involves individuals. People change only insofar as they see, know, and are aware of reality. This book is an attempt to reflect reality, as science reveals it, so that people can learn from it. My hope is that we can be considerate of the quality of life for the future generations of all species, for the future for all forms of life. However, my advocacy puts me at risk, appearing as though I know the details about what is next, how to implement such management, or what to do now. These are things I cannot know any more than anyone else, although I hope that I have something to contribute. I have tried my best in this book to provide a way to get guidance for our goals as a species consistent with goals as individuals, communities, and societies in all our interactions with both each other and the nonhuman—sustainability at all levels and in all realms. Although there is management advice found in the information I have displayed, this book is much more a prescription for finding that kind of advice—learning from

and mimicking nature. As in our personal lives, it seems that the best way forward for our species is to proceed based on what can be learned from collective experience, now including the experience of other species.

Management itself, however, involves many details, determined in part through the resolution of objectives for our species in a way that reverts to us as individuals. Systemic management leads to its own implementation and involves many things we already know how to do quite well,[1] amplified by learning to do things we have never done before, discontinuing doing some things we are doing now, and doing many things differently.

Systemic management is a reality-based self-guiding process. When we manage our personal lives, we rely heavily on experience—largely in a systemic way (Zimberoff and Hartman 2001; or as a trial-and-error processes in the comedy of survival, Meeker 1997). Our lessons as individuals mean a great deal to us and we find it difficult to avoid making personal needs pre-eminent when it comes to deciding what to do as a species; at that level, species-level experience is necessary and may be the opposite of what individual level experience would have us believe. What is good for us as a species, may not be good for ecosystems or the biosphere—what is good for a part can be lethal for the whole. Fatal flaws can evolve (Potter 1990). Thus, to define large-scale objectives, I had to acknowledge my own limitations, especially my ignorance; there is something to be said for the wisdom of insecurity (Watts 1951). I am not a species and cannot have that experience. At all levels, guiding information is found in nature and experience. However, when species-level goals and objectives are established, there are subsidiary issues that ultimately break down to action by individuals. The Achilles heel of conventional management seems to be the thought processes, belief systems, and values involved in establishing policy—they tend to be the basis for management. In contrast, clear thinking is fundamental to asking the right questions (Chapter 5) in systemic management and empirical information becomes the basis for decision making to replace the human constructs that make conventional management so prone to error. Thinking is crucial in directing the science needed to observe guiding information, but the guiding information is ultimately reality-based (systemic). Guidance based on the alchemy of converting nonconsonant information to goals and policy (rather than using such information to reveal correlative patterns) brings on more problems than are solved.

Thus, seeking guiding information that eliminates gaps between science and management is a key element in systemic management, and depends upon asking the right questions (both for defining goals and guidance in achieving them—fractal, involving all levels of biological organization). However, large-scale questions and management action do not find resolution if individual-level elements are ignored. The personal, interpersonal, and psychological[2] dimensions that are involved are beyond the realm of this book partly because of their complexity. This is not quite the problem it might seem on the surface, however, as our thinking, habits, and actions (and their effects from the past) are accounted for *a priori* in integrative patterns—we are part of reality. Trial-and-error processes are universal phenomena, including those through which we have learned a great deal regarding personal or individual matters. They are basis for change and changes in our thinking and beliefs are a key part of systemic management.

There is an extremely superficial quality to the few objectives for systemic management found in this book. I have been able to present only a tiny bit of information regarding goals for our species' relationships with other natural systems in their hierarchical organization (e.g., our interactions with, and influence on, other species, ecosystems, and the biosphere). My treatment of implementation is even more superficial. Even though systemic management will provide its own guidance, it cannot happen without the openness, and resolve, to actually make the changes that are revealed as necessary by empirical patterns. The examples I have presented are meant to serve as models of the process, to initiate the adoption of systemic management, and to help point to the observing (and changes in thinking) that can circumvent current impasses and be the basis for hope. The questions initiated for the eastern Bering Sea represent only the most minute of steps toward implementation for only one example. The majority of the

preceding chapters dealt with the rationale behind systemic management, a few examples of goals to be achieved, the methodology for refining goals, and opening the door to more detailed information regarding implementation. However, any realistic grasp of the potential brings the realization that we are faced with a daunting, possibly impossible, challenge.

Possibly impossible? I do not want to underestimate the difficulty we face. Nor do I want to appear idealistic in the face of the reality of that difficulty. The problems before us are immense and current efforts to solve them are extremely superficial. The misapprehension of the magnitude of our problems is exemplified by current goals for reducing CO_2 production—reductions of 50% or 80% fall woefully short of the mark and would have to be repeated several times to achieve true sustainability. Nevertheless, there is progress: many people are aware of global warming and our contribution to it. Of more importance, is the realization that debate over such issues is a matter of opinion that can lead to very useful management questions. However, the vast majority of problems like global warming are not given similar attention by worldwide team efforts to educate people. Thus, as I am often reminded, it is highly likely that the adoption and implementation of systemic management is beyond our capacity to accomplish. I am not sure which is the most difficult: accepting the systemic approach or making the changes the approach requires. To succeed, both will have to happen, but both present immense challenges. In fact, if I were to wager on the odds of our achieving sustainability, I would bet against it. Why? What is the basis for considering the chances of success so slim?

Acceptance and implementation both involve individuals directly. This simultaneously exposes both the hope and the extreme difficulty. Implementation will clearly be a monumental undertaking. In view of the difficulty, reduction of the human population, for example, may occur only as a result of the nonhuman forces of nature rather than the result of any direct form of management. Cutting our harvests of fish by 90–99% seems draconian in conventional thinking as does the reduction of our energy consumption by 99.9%. The problems we face may be characterized as a more generic form of nature deficit disorder than envisioned initially—natural selection and natural limits have not been allowed to have their normal effects on our species for centuries and we experience being extinction prone as a serious risk yet to be acknowledged. However, acceptance of systemic management may be even more demanding than implementation.

A crucial step is a systemic change in thinking. If there is anything to what I have presented in the foregoing chapters, systemic thinking largely removes us from the process of decision making when it comes to establishing specific goals for our species—a humiliating thought (humility has long been recognized as essential; Bateson 1972, Jacobson 1990, McIntyre 1998, Meffe *et al.* 2006, Rasmussen 1996, H. Smith 1994, White 1967). Many scientists will find it nearly impossible to back out of their role as advisors and consultants to focus, instead, on asking good management questions—questions that then define the science they can conduct with results they can convey to managers as self-evident advice. The role of scientists as experts becomes confined largely to being specialists regarding patterns consonant with specific management questions. This takes advantage of realistic reductionism to avoid the misdirected reductionism (Belgrano and Fowler 2008) of conventional management.

The complexity and reality of the human system presents what may be insurmountable obstacles nearly impossible to overcome. Maybe it is impossible; the strong possibility of our being on the path to an evolutionary dead-end is very real. We are subject to the pruning rules among species; we are, after all, a species. Nature will take its course whether or not we change our thinking. Although we are witness to an anthropogenic systemic shock wave comparable to those of historic meteor strikes, Mother Nature recovered from those and is likely to do the same after our demise (or population reduction). As such, we have the option of doing no more than watching ourselves as victims of what we are, and what we do—results of the forces of nature around and within us. As Meeker (1997) says: "Humanity may have to settle for the distinction of being the first species ever to understand the causes of its own extinction".

It may be our only option. Much of what we are and do is beyond our control. It is entirely possible that we are beyond achieving true sustainability. We experience our genetic programming for procreation, our evolved dependencies on other species (e.g., clothing, drugs, food, fiber), our emotions concerning death, the pangs of hunger, and empathy for other humans as they experience the aftermath of our past. Recreational activities that are destructive to nature seem to us to be essential; they are taken for granted as realistic options. We think we understand human rights and hold them sacrosanct. A similar hubris is behind our economic systems. Economic values trump ecological realities in the majority of conventional decision making. These things serve us fairly well as individuals or even as human communities. However, as a species, many such ways add to the risks we face; the extinction of humans involves the end of all our economic systems, business, and industry. The way out may be impossible owing to what we have become. Many of the reactions to this book can be expected to add to the list of ways it can be seen as impossible—including the many reactions starting with the words: "Yes, but...". The phrase "Yes, but..." is usually the introduction of an argument to stay on course,[3] a course called into serious question when we open our eyes. It is a course of misdirected reductionism (Belgrano and Fowler 2008) for which the fatal errors have been recognized for thousands of years. Why should we expect things to change now?

As a species becoming self-aware, we may be much like the adolescent faced with two choices: (1) taking advice from others based on their experience or (2) learning from personal experience (occasionally fatal). Although we may be quite adolescent as a species (Plotkin 2008), this choice, of course, is not confined to adolescence. In the case of being a species, we face this choice in learning from other species and history—natural history. The experiences of historical societies constitute a history from which we can learn pertinent lessons (Costanza 1995, Costanza *et al.* 2005, Diamond 2004, Ehrlich and Ehrlich 1996, Ponting 1991, Redman 1999). We fail to learn from history so frequently that it is easy to argue that it will not happen now—the fatality here being our extinction.

As much as I expect that it will not happen, and as difficult as it will be, I remain convinced that systemic management is a workable option;[4] otherwise I would not have persisted in publishing this book. There are examples where we do learn from others and from history; we know we can. Although world war, starvation, and disease of monumentally abnormal proportions, may lead to our own extinction, as a result of our past management practices, I believe we can take meaningful actions that deal with the complexity of being sustainable, including these risks. At least four factors provide hope:

- immense change can be accomplished with time,
- humans are adaptable and resilient,
- goals and objectives are helpful,
- there is power in the human mind.

This list assumes that those cases where we *have* learned from the experience of others, and from history, are experiences about the learning process from which we cannot only benefit from learning but also learn about learning. In the latter, the learning I refer to is finding that we can, in fact, learn this way. The lesson is a meta-level lesson; we are not simply learning this way but learning that it is a way to learn. It is thus categorical learning, fractally inclusive, hierarchically transcendent learning, or Bateson's (1972) deutero-learning—an experiential learning about learning. In the process, we know that it is the pattern that is important. The advice of an elder or mentor, is not the same as the common ground of the advice of many. What we can learn from other species depends on patterns—information from numerous species.

Change will take time: The changes required of us are not to be made overnight. While some changes (like changing our minds) can conceivably take only years or decades, others will take a long time, even hundreds to thousands of years. Abnormally extreme measures are not part of systemic management unless they are precipitated by systemic reactions (either as reactions to our achieving the normal in ways where we can, or in reaction to our lack of action). Purposeful individual sacrifice that is abnormal would be contrary to the spirit of systemic thinking but is arguably better than "staying the course". However, this is not a rationale for procrastination—the sooner we get started, the better. Nature may "beat us to the draw" with a set of what we will consider catastrophic, and even abnormal (infrequent, but entirely natural,[5] and, in the long-term, normal) adjustments in the organizational dynamics of the systems of which we are a part. A reaction by ecosystems or the biosphere to result in a pandemic or our extinction, under current circumstances, would be perfectly natural as well as normal.

Humans are adaptable and resilient: Further basis for hope is found in our adaptability. Our history of adaptation brings along with it the recognition that the changes we need to make are within reach. Change is often spurred by the depths of hopelessness—there can be hope in hopelessness! We are a very resilient species, owing, in part, to the resilience of individuals. Some individuals are more resilient than others. Although we may have brought ourselves to a situation where a great deal of suffering and death may result (and be necessary), our resilience is basis for hoping that we can change to avoid more extinction, especially our own. Even more optimistically, our resilience is basis for hoping that we can do more than avoid past errors: it can be used to achieve sustainability.

Goals and objectives are helpful: Hope can also be found in accepting goals and objectives. Hopelessness would seem to be a natural byproduct of aimless activity in the face of problems with no obvious solution. With the goals and objectives that emerge through empirical information, a sense of direction is basis for hope. Hope can be engendered by the communal spirit of common direction. Social facilitation of progress is embodied in prestige given to those who take steps in the direction of achieving systemic goals. Any actual progress toward those goals is quite meaningful and serves as the basis for hope, as tiny as the steps might seem—problems get created incrementally and they can be solved incrementally. To the extent that we become aware of the magnitude of problems, progress is made toward the goal of an informed public and such progress engenders hope. There is hope that an informed public will result in the emergence of unpredictable, but creative, approaches to achieving empirically defined objectives.

There is power in the human mind: Finally, the nature of our minds also offers hope. Our intellect or intelligence may make us a novelty in at least one way that can work to our advantage. Novelty is a part of the trial-and-error dynamics of natural selection and emergence; some things work—intelligence may be one of them. Intellect is a double-edged sword, however, and often serves to draw us into the problems we now can measure for our species (e.g., see Rees 2002). Our intelligence can be, and is, used to deal with the short-term matters so familiar to everyone—misdirected reductionism (Belgrano and Fowler 2008). As Redman (1999) points out repeatedly, decisions by previous societies seemed logical at the time, but ultimately led to their demise (see also: Costanza 1995, Costanza *et al. 2007,* Diamond 2004, Ehrlich and Ehrlich 1996, Ponting 1991). Likewise, our intelligence could become another of the species-level characteristics that leads us to extinction[6] along with most other species. Human nature has resulted in the variety of wonders we have accomplished, including those extremes exemplified in Chapter 6. Our species may merit being listed in the *Guinness Book of World Records* simply for the number of times it *could* be listed—a meta-record.

Nevertheless, the human mind also provides the means to recognize patterns, problems that occur in patterns, lessons from the past, and lessons to be learned from observing nature to face and deal with larger scale problems. Recognizing problems caused by our own limitations, particularly our thinking, is a step forward. We can draw upon our capacity to wisely sort through our options, distinguishing between those that will work and those that will not. A major message of this book is that, following Nash's philosophy, things have to work at all levels; as Carse maintains, they have to be part of an infinite game. Less tragedy and more comedy (Meeker 1997) are options at the species level. In the spirit of adaptive management, nature has supplied us with answers to our questions; biomimicry is an option at all levels. Collaborative learning (e.g., Clinchy *et al.* 1996) can extend to the level of the species; other species may not be able to learn from each other's experience, but we can—we can

be mentored as a species by other species. We can see and use the information in nature's Bayesian integrations of reality. We can use our intellect to see and mimic examples of things that work, and know what constitutes reliable guidance in full consideration of complexity.

I have tried to present the content and arguments of this book in an objective manner. I join others in my failure in this regard. My principal bias stems from care, concern, and consideration for the life-systems of our world, including our species and its long-term chances of survival—future generations and the well-being of individuals in the process of change. While humans are of concern, the well-being of all other species, the individuals of those species, and the living systems of which they (and we) are a part are all at stake. These include ecosystems and the biosphere. In order to be sustainable, we have to be in mutually sustainable relationships with each other—in part, the lessons of ecology. What happens will be influenced by what we do, regardless of what we do. In view of the ways Mother Nature works, it seems that we have to ask ourselves if there is justification for change, if we want to change, and then if we have what it takes to do so.

This leads to two clearly subjective issues that seem important:

1. Elements of change—what are some of the underlying changes that might lead to adopting systemic management?
2. Human factors—what are some of the human institutions and their roles in management and change?

7.1 Elements of change

It has been my experience that our current belief systems are, themselves, major obstacles to accepting reality as the basis for management. One of the most effective elements in our path around these obstacles would seem to be education. In the spirit of systemic management, I think that we need to be reminded that education may be more a matter of experience than attempts to learn in structured classroom settings. As Eleanore Roosevelt said: "Learn from the mistakes of others. You can't live long enough to make them all yourself" (Roosevelt 2002). This applies at the species level. Once we have arrived at beliefs and understanding consonant with reality, we will be able to move beyond words to management in action. This section considers each of these in turn:

- Belief systems that stand in our road
- Education as an agent of change
- Management in action.

7.1.1 Belief systems that stand in our road

Belief systems are part of human complexity and their influence from the past contribute to what we see around us—are accounted for in natural patterns, especially those that expose our abnormality. One of the beliefs that has contributed to our failures is our belief that so many of our models (concepts, information, or knowledge) of one or more thing(s) can be used without correlative conversion in the management of something else. This involves serious "disconnects" and results in misguidance in the translation of information to artificially make the connection. Our choice of information rarely fully informs management. We fail to recognize the importance of achieving full consonance between management question and guiding information. Models involving/about related factors are only that; they are insightful and help understand. They are not the emergent patterns showing us what actually works. Models that are most useful are those that represent patterns consonant with management questions even if the representation is imperfect. The patterns are reductionistic in their specificity, but reflect the reality involved in their emergence; the reductionistic nature of science becomes its strength—at least we can take advantage of this limitation of science to achieve the holism required of management.

During the 1970s, I taught courses on how to achieve maximum sustainable yield (MSY) in fisheries (and wildlife) management at Utah State University. Since then, I have experienced a growing clarity about both the utility and the danger of such approaches—models are indispensable analytical tools for better understanding, and measuring, for discovering and representing patterns,

and processes, but *are not the reality they attempt to represent*. Like words and concepts, they are not what they represent (Watts 1951); they are abstractions (Pilkey and Pilkey-Jarvis 2007). More importantly, they are rarely in the units, or of the logical type, framed by our management questions when we take the time and effort to be clear about what those questions are.

We humans show a remarkable tenacity to believe that we have the skill and capacity to reconstruct realities in models, concepts, or theories. This runs contrary to all experience of reality wherein every perception is only a representation of a part of reality. Each is a reductionistic depiction, not complete ultimate reality. This is true, whether in the various disciplines of science, politics, or in metaphorical stories of the problem itself (e.g., the blind person examining an elephant). The Humpty Dumpty phenomenon mentioned several times earlier in this book always prevails, whether it involves an individual or a group of people. The belief that we can piece things together to work out solutions, rather than observe the solutions in nature, always prevents a full accounting of complexity. To me, it seems all too likely that our tendency to think otherwise is one of the results of our abnormal connections with nature (Appendix 1.2, a symptom of nature-deficit disorder, Louv 2005) such that our thinking is another victim. In the words of Thomas Berry (Berry 1997): "If we devastate or distort the natural world, to that extent our inner world is distorted because our inner world is determined by our outer experience". Through connections we have in today's world (including through scientific observation) we are victims of systems abnormally altered by our effects on them—a serious "catch 22" (Appendix 1.2).

Wilson (1998) claims that one day we will have produced enough knowledge in scientific endeavor to achieve a full accounting of complexity, or at least an accounting adequate for decision making. Wilson's belief is typical and is admirable and good as far as it goes. However, complexity involves the infinite—ultimate reality (Appendix 1.1)—and no man-made combination of the finite can ever completely recreate this reality (nor fully represent it, Bateson 1972, Watts 1951). Models are always based on subsets of reality (Caughley 1981, Pilkey and

within patterns consonant with their management questions, whether such variation is among individuals, species, or ecosystems. For species this would include CO_2 production, size of parks and another protected areas, energy consumption, resource consumption, and population size (including density and distribution). Research to examine the limits to variation in relation to temporal and spacial heterogeneity, would be funded and encouraged as a form of science preferred to research that makes obvious the fact that reality is complicated. Academic science cannot be dropped as it helps lead to more and critical management questions; it helps identify correlative patterns to make direct use of the information scientists produce (there is a huge difference between correlative conversion of information and the alchemy of conventional conversion).

Governments, along with the rest of society, would support educational programs to make people aware of cases wherein humans are abnormal and what needs to be done. Stress would be placed on the unity of purpose that would emerge among governments, businesses, the sciences, religion, social and environmental organizations, and other elements of society as forced by the consistency of nature (Hobbs and Fowler 2008). The reality of such consistency is emphasized experientially in the consistency of problems we see in human abnormality.

Environmental organizations, for example, would not give up on their missions. However, goals and objectives would be consistent among organizations, whether they are concerned about human population size, resource harvesting, pollution, immigration, habitat, biodiversity, extinction, CO_2 production, or other elements of reality. The production of more energy (using wind, tides, sunlight, garbage) is inconsistent with a pattern showing that energy usage is already abnormal. The goals and objectives for such organizations would be consistent with those of government. There would be no more reason for difference of opinion than there would be among physiologists, geneticists, morphologists, and neurologists regarding the appropriate body temperature for individual humans, hummingbirds, or elephants. To the extent that achieving systemic sustainability is viewed as utopian (or proves in fact to be impossible for us to achieve) we have grounds for being skeptical about success.

7.2.4 Religions

At their core (as opposed to many or most institutional and personal manifestations), the various wisdom traditions of the world place primary importance on the concept of veracity (truth or truthfulness, H. Smith 1994). Facing reality/truth is at the core of systemic management and is a common element of both religion and science; systemic management is reality-based to include any God or gods and beliefs about them. Religions can debate about who are the most believable and historically recognized messengers regarding reality, much as science does with its prominent contributors. However, both are dealing with the underlying assumption, or belief, that there is a reality to be understood, observed, and accounted for in decision making. Both face the fact that the whole of reality is made up of its constituent parts—the realities we face in day-to-day life. The holism of religion and the reductionism of science meet and find common ground in systemic thinking and the consonant reality-based aspect of systemic management. Science and institutional religion may share major imperfections owing to human limitations—so many as to make comparison of religion and science as practiced today a ludicrous comparison. However, we can see that we are "throwing the baby out with the bath water" if the common elements of reality behind them are ignored—especially in terms of changes in both that can reclaim consistency.

Movements are afoot in spiritual and religious circles to deal with the problems that we face in the more inclusive living systems of the world of which we are a part (Rasmussen 1996, Swimme and Berry 1994). These are exemplified by organizations such as Earth Ministry, National Religious Partnership for the Environment, North American Coalition on Religion and Ecology, Interfaith Network for Earth Concerns, Interfaith Center for Environmental Stewardship, Coalition on the Environment in Jewish Life, and the Evangelical Environmental Network. Systemic management points the way for such efforts to recognize unity of purpose through the common ground of reality (i.e., the infinite of Fig. 1.4, Appendix 1.1)—not only among such organizations but also between such organizations and other institutions, governments and organizations.

Religion and science may find relief from polarized perceptions if open-mindedness in both realms finds common ground, not only in the immanence of God as natural law (Morowitz 2002) as well as in other *parts* of reality, but also in the transcendence involved in the *emergent* nature of everything. Humans are not the only species to emerge from reality. The larger (more inclusive) wholes seen in ecosystems and the biosphere are subject to pruning rules (including natural selection among species; Chapters 3) acting on their parts to give rise to emergent qualities also. If the pattern of emergent qualities applies to everything with parts, the universe must have its own emergent qualities. The composite of such qualities at all levels is a form of transcendence. Reality has both an immanent and a transcendent nature simultaneously. As many have noted, science and religion may have much more in common than one would be led to believe in watching and listening to the superficial of popular discourse.

Religion, more than science, assumes the role of defining morals, ethics, and the right or wrong of what we do. Evil, as defined by Reinhold Niebuhr (Brown 1986), involves action without regard to other parts of the universe around us. It is this perspective that is behind the concept of systemic management as thinking and action that account not only for the nonhuman but the human on time scales heretofore largely ignored. Consideration of the nonhuman is inherent to asking questions such as "How much primary production should we leave for other species?" Patterns in what is left for other species, by other species, provide integrative holistic bases for addressing such questions. Action based on such information involves compassion and justice well beyond anything achieved in the largely anthropocentric approaches of today, while avoiding the opposite extreme of biocentrism. Consideration of the nonhuman is inherent to asking all management questions so as to seek and use consonant emergent information.

7.2.5 Individuals

Much of what we do as a species is done through the collective action of individuals.[24] This places the responsibility for success where it brings out the greatest difficulty as well as both the greatest control and potential. Sacrifice, effort, and cost[25] are involved in the actions individual humans must take to bring about the changes needed for our species to achieve sustainability. To the degree that we experience the conflict of seeking personal good when it runs counter to what is necessary for the sustainability of all species (and ecosystems and the biosphere), we are experiencing responsibility not recognized by previous generations. Reducing birth rates (for those who value large families), consumption, energy use, CO_2 production, and all the influence we have on the ecosystems and the biosphere will reduce our quality of life as defined by current (especially western European/North American) standards. This predicament can only get worse for future generations if we do not make sacrifices in taking action now. Sufficient changes now (or in the near future) will increase the quality of life for future generations (of both humans and nonhumans); indeed it can be expected to enhance the likelihood that humans can continue into the long-term future. In the long-term, a reduced human population can be expected to result in less rather than more international conflict over resources, starvation, malnutrition, and the risks of disease (including evolution of new or more virulent diseases)—if ecosystems and the biosphere have time to recover from changes we have caused, and are causing, so as to support us.

Huge challenges are ahead in reducing the risks we face. These challenges include the scope of change required of individual humans—including changes in our consciousness, thinking, habits, and values. The momentum of change in the opposite direction has been building for centuries. We need to remember that natural selection at the individual level often results in factors that are not adaptive at the species level (evolutionary suicide, Combinations 5 and 6, Table 3.1)—cases where what is good for the part is lethal for the whole. Thus, emotions and values (especially those based on monetary economies) can easily be in conflict with natural forces involving greater complexity. Unless we can overcome our individual genetic programming (along with the socially and culturally learned programming) and consciously act for the benefit of species-level survival, the consequences will include increased risks of starvation, plagues, war, and ultimately our extinction. We can act in our choice of leaders, in the organizations that we contribute to, in our selection of mates, in our use of resources, and in the number of children we have. Whatever we do, however, it will not ultimately work if it is not to the benefit of both individuals and our species, in the spirit of Nash equilibria (Combination 1, Table 3.1) or infinite games wherein the objective is to continue the game (Carse 1986).

Individuals can ask questions. For example, managers can be asked if the evolutionary impact of any particular management strategy has been taken into consideration. Has the decision to permit the development of land (e.g., for recreation or shelter) accounted for evolutionary impacts on other species? Is a sustainable level of energy, space, or biomass being left for other species—not just one, but *all* other species? Have managers taken into account the degree to which ecosystems will be altered and how such alteration results in feedback to our species? If so, how were such considerations weighed against economic, social, or other human interests without those interests superseding sustainability? Were decisions made so as to account for future generations of both humans and other species? Has the establishment of harvest quotas, limits, or zones for harvesting resource species (e.g., fish, trees, or commercial crops), taken into account the evolutionary impact such takes have on both target species and other species? Have you taken into account the monoculture effects of modern agriculture? Have the butterfly effects of any particular action been taken into account? The impossibility of such considerations in conventional thinking will hopefully be acknowledged so as to open minds to adopting systemic management and then be able to answer in the affirmative.

Individuals can write, publish, talk, and form action/advocacy organizations.[26] Books (including novels), plays, art, educational materials, and advocacy organizations aimed at each of the species-level dimensions for which we find ourselves to be abnormal would be good steps toward making the human population aware of the magnitude of our problem(s). The mass media may be one form of technology that serves us well if used to educate people around the world of the plight before us.

Individuals are involved in politics, religions, education, and other social institutions. I agree with Brian Czech (2000) who sees major social change as a key element in achieving sustainability. It cannot be done with a change in economic thinking alone. It will have to involve our legal system. It will have to involve the arts, the media, and much more. It is not clear that it can happen. It is clear that it will be impossible without motivated individuals armed with an understanding and appreciation well beyond that involved in the decision making and management we witness today. The changes must be systemic.

7.3 Looking back and looking forward

The preceding chapters bring us to the realization that the problems before us are immense—much larger than acknowledged in conventional thinking,[27] in fact, larger by orders of magnitude. Solving these problems is systemic management. Changing is action (praxis) that is reality-based. It is a holistic form of management with parts exemplified by ecosystem-based components, biosphere-based components, and single-species-based components. Science provides us with measures of the human-based problems, as problems open to solution through human action. Science provides us with explanation of these problems, including an understanding of the concept of emergence and the complexity of factors involved (Plate 7.2). Science helps us understand the integrative nature of guiding patterns so that their use in establishing goals for management account for the infinite nature of reality.

However, there remains the matter of action to achieve such goals. What can or should we do to achieve the objectives of sustainability? The means are different from the ends. We are at the point of realizing that the importance of achieving the goals is challenged only by the importance of how we achieve them. In the spirit of systemic thinking, are there "patterns of praxis" that inform us in this regard? Can we find information in the nonhuman that tells us about what to do to achieve the goals we find from the same source? If not, can we use our own experience?

We can carefully use examples of success in current management action—now with realistic goals. We have legislation in place for, and organizations with experience in, setting quotas for the harvest of marine fish resources—quotas can be set to avoid the abnormal. We regulate the harvest of timber, and the occupation of land—those processes can now be used to avoid the abnormal. Successful means of regulating ourselves can be carried forward by replacing existing goals with the more realistic goals established by integrative patterns consonant with the management question involved.

Caution, however, is imperative. It is tempting, for example, to take information such as that provided by Pimentel *et al.* (2007) and move toward producing more food, exercise more disease prevention, distribute food more equitably and seek/develop technology to help. We know how to do such things and we know that there are problems associated with technology (Pratarelli and Chiarelli 2007). We now know that the use of technology exacerbates problems such as energy and resource consumption by our species at rates that are already pathological. Such action would not solve the problem of excessive human numbers, the abnormal production of CO_2, or production of chemicals (some of which are behind the problems outlined by Pimentel *et al.* 2007). Research exemplified by Pimentel *et al.* (2007) is exemplary of the basis for asking clear management questions. Again, the patterns produced by science in response to good management questions might lead us to think that the matter of change is simple: we know how to achieve goals when they are clear.

As with all things, its not quite so simple and we begin to realize the magnitude of the challenge before us. By direct action to achieve the goals made clear in the preceding chapters, we immediately encounter the problems that we have not been able to deal with in the past—problems such as our overpopulation. Now we can use systemic management to cut back on our consumption of energy, our production of CO_2, our occupation of space, our practice of medicine, agricultural production, use of chemicals, and the multitude of other ways that involve human excess. In most, if not all cases, there would always be the consistency between such actions and associated results involving a decrease in our population through starvation, disease, and other factors that are often judged as draconian (and unacceptable in current thinking) when we face their reality. Systemically, this would be an approach that would reduce most if not all of the pathological effects we have on the world around us and solve the problem of our overpopulation, along with all associated abnormalities.

By doing so, we directly encounter the emotions and human values that prevent objectivity in conventional management.

Such an approach solves more problems than are being solved conventionally. It can be seen as an approach in which human intellect finds a means to avoid an abnormal risk of human extinction—an evolutionary feature sufficient to transcend itself and realize its own dangers. If this novelty (like the repeated evolution of the eye) has species-level advantages in species-level fitness, maybe it is enough to work with. It would be a matter of an infinite game (Carse 1986) in which we humans took the initiative, acted proactively, took responsibility, and faced reality. In terms of an infinite game it would involve one of the currencies of ecology—life and death. Systemic management is not a matter of simply sending a check to one's favorite environmental organization.

Certainly, human values, when we let them be our guide, lead to the quick conclusion that death is an unacceptable means of achieving sustainability. Human values and the risk they present in the form of a fatal flaw leading to evolutionary suicide, however, are to be reckoned with as parts of the reality we face. They put us in the position of remaining where we find ourselves now—using human values as the basis for management (top row of Fig. 1.1), rather than the transcendent value of sustainability as it applies to all systems, human and nonhuman.

If we follow the pattern of using our values to direct us to the task of asking management questions (rather than set policy), we see another option. Are there patterns in praxis? What do other species do to find their place in natural patterns? What we find, when we look to other species, is that very few, if any, species exercise self-restraint as a species. There are biosocial factors that contribute to population regulation, for example, but intrinsic (endogenous) factors, for the most part, are not willful other than that they serve for the fitness of individuals and incidentally work to regulate population (Fowler 1995). If gorillas had evolved the capacity for tool use, medicine, and agriculture, instead of humans, it would probably be they, rather than us, faced with the dilemmas we now confront. Comparisons of historical human cultures brings us to the same conclusion regarding subpopulations of our species (Costanza 1995, Costanza *et al.* 2007, Diamond 2004, Ehrlich and Ehrlich 1996, Ponting 1991, Redman 1999). What we find is that it is the organizational dynamics of the systems of which we are a part that are primary in giving rise to patterns by placing limits on their components (Ahl and Allen 1996). We, along with other species, are limited by the systems of which we are a part. There is only a finite amount of space, a finite amount of energy, and a finite amount of each elemental resource (e.g., phosphorus, sulphur, chlorine) to be shared among the species that occupy our planet. The effects of the collective set of species on any one are more powerful than the effect of any one on the others (noting the current effects of humans as a dangerous potential, but temporary, counterexample).

Where are we led with this information? Our best option may be to accept the forces of nature and let them do the job of restoring humans to a normal fit within the universe. In this regard, management action would be a matter of letting nature's forces give rise to the mortality and reduced birth rates necessary for the infinite game to be played realistically. This would promote a natural form of selectivity, rather than unilateral action by a world leader (*a la* Hitler or Pol Pot), or even a democratically adopted regime for population reduction (all involving politics). Such a regime would, itself, violate the principles of systemic management in addition to being unethical by current standards

(here we see an example of human values that are consistent with systemic action—not always the case). As difficult as it might be to accept, and as horrendous as we might evaluate it to be, the acceptance of nature's forces, may be our best option. We can hope that they are not so extreme as to result in our extinction—perhaps sparing a few indigenous cultures that are capable of living sustainably. We can realize that the unintended consequences of medicine and agriculture, technology, and conventionally applied science, as products of an evolving mind, are among the reasons we find ourselves in the current predicament. They serve well in the interest of individuals, but not our species.[28] Our beliefs and modes of thinking are involved and have their consequences. As much as we fear diseases, death, and starvation, acting on such fears is an example of cerebral alchemy: translating an emotion into action rather than using it to turn to the process of asking good management questions.

We seem to be faced with two options: continuing on the course we are now taking (basically making no change, in which case the forces of nature will do the job for us), or taking action ourselves in accordance with the laws of nature (in which case the forces of nature also work with us). The former involves more risk, with increasing risks as we move further into the territory of the abnormal or pathological. The latter makes use of intellect in a way that undoubtedly has no precedent in evolutionary history; we have a species-level self-awareness that reflects back on our self-awareness as individuals. The latter, more than the former, involves an encounter with reality akin to a mystical experience. There is a choice.

7.4 Conclusion

Researching and writing this book have involved personal lessons beyond my wildest dreams or expectations. Informed individual action to change our species seems to be a critical element in our achieving effective management. We need leadership and guidance to help achieve sustainability, and individual motivation and action is fundamental. We need realistic guidance—based on reality. The inability of humans to control the natural world in all its complexity—individuals, species, ecosystems, and the entire biosphere—is apparent. Giving up our tendency to attempt control of the uncontrollable requires a change in paradigm regarding our perception of reality—the same change that leads us to guiding information. This change goes beyond the rational to include the emotional, aesthetic, and spiritual and depends on experience. I sense that we are just beginning to perceive ourselves as a species facing a situation much like individuals do in their personal lives. While, as an individual, I cannot orchestrate my entire world, I seek to adapt myself to live as well as I can in circumstances largely beyond my control. The main thing under my control is myself, and I would be fooling myself to believe I have complete control there. In accord with elements of eastern philosophy, trying to control ecosystems or the biosphere is self-defeating.[29] Based on an understanding of life[30] augmented by the concept of natural selection at multiple levels, the best we can do as a species is to make change with, and be guided by, empirical patterns in nature—the infinite guiding the finite.

Notes

Chapter 1

1. Sustainability is defined for the purpose of this book as "what works". It is the normal, as opposed to the abnormal (which does not work). It is the healthy in contrast to the pathological; it is the genuine compared to the artificial, or abstract. See Fowler and Hobbs (2002) regarding the importance previous efforts have placed on recognizing limits in defining what is normal for management.

2. Emergence, as used in this book, means that there is an explanation to the origin of everything (Belgrano and Fowler 2008, Morowitz 2002). We cannot know the complete story of this origin but it involves all contributing factors (see Bateson 1979, Clayton 2004, Emlen *et al.* 1998, Jørgensen *et al.* 1999, Lewin 1992, Nagel 1979, Nielsen 2000, Prigogine 1978, Scheiner *et al.* 1993, Solé and Bascompte 2006), and these factors (reality, Appendix 1.1) are accounted for in what we see.

3. Consideration, mention, or concern about human extinction as the result of human-caused changes (feedback from our influence) is frequently found in the writings of biologists (e.g., Boulter 2002) and environmental scientists. But it is also found in the writings of a variety of other philosophers and scientists. Examples of each include Bateson (1979), Brown (1995), Burns *et al.* (1991), Cairns (1991), Capra (1982), Catton (1980), Cawte (1978), Chaloner and Hallam (1989), Chiras (1992), Darwin (1953), Ehrlich (1986, 1989), Eldredge (1991), Garrett (1994), Greenway (1995), Hassan (1981), Hern (1993), Jarvis (1978), Jenkins (1985), Jørgensen (1992), Laughlin and Brady (1978), Lederberg (1973, 1988, 1993), Macy (1995), McNeill (1989), Mines (1971), Noss and Cooperrider (1994), Ovington (1975), B. Patten (1991), Pennycuick (1992), Ponting (1991), Potter (1990), Reed (1989), Rosenzweig (1974), Roszak (1992), Salzman (1994), Tiger and Fox (1989), Trotter and McCulloch (1984), Tudge (1989), and Whitmore (1980).

4. The value of ecosystems, in human terms, has received extensive treatment in recent years. This work is exemplified by Ehrlich (1991), Ehrlich and Mooney (1983), Ehrlich and Wilson (1991), Farnworth and Golley (1974), Myers (1985, 1989), Nash (1982), Nash (1991), Norse (1985), Norton (1987, 1989), Rapport (1992), Sagoff (1992), and Seal (1985).

5. A discipline of psychology called ecopsychology focuses on such issues at the personal level (Roszak *et al.* 1995). Other aspects of separation from nature are likely more emergent at the species level. A number of authors have treated these issues (e.g., Bateson 1972, 1979, Bateson and Bateson 1987, Gore 1992). The issue of control and our attempts to control may be one facet of potential pathology in these matters (Holling and Meffe 1996).

6. Such experience may lead to comprehension of nature that is occasionally seen as an alternative (or compliment) to understanding gained through western science, which is often subject to criticisms by scientists themselves. These criticisms often center on the mechanistic views that have failed to produce the understanding and explanations once thought to be so promising of Newtonian–Cartesian based science (see Mayr 1982, Peters 1991, Rosenberg 1985, Simberloff 1980).

Comprehension involves perception and fundamental to perception is experience (Wilber 1996). Simple communication involves personal experience of this principle. Communication with a person who has no personal experience with the subject matter is nearly impossible (certainly more difficult) when compared to the mutual understanding between two people both of whom have common personal experience. People who live in proximity to nature (within ecosystems, in direct contact with the sources of their food, sources of materials, and the limitations of diseases, predators, and competition) have perceptions and comprehensions not possible for those that learn secondhand from classes, reading, explanations on television, or even research.

7. For an introduction to problems and limitations identified in the use of models in ecosystem studies see Gross (1989), Hagen (1992), Hedgpeth (1977), Holling (1993), Kareiva (1989), Karplus (1977), Kingsland (1985), Moran (1984), Orians (1974, 1975), Pimentel (1966), Pimm (1991), Schlesinger (1989), Schnute and Richards (2001), and Ulanowicz (1989). For a treatment of the more general limitations of models see Pilkey and Pilkey-Jarvis (2007).

8. There are many references documenting, listing, or describing specific laws or agreements that require management that includes ecosystems (e.g., Angermeier and Karr 1994, Belsky 1995, Christensen *et al.* 1995, Clark and Zaunbrecher 1987, Davis and Simon 1994a, Francis 1993, Jarvis 1978, Keiter 1988, Mannion 1991, Munn 1993, Noss 1990, Rapport *et al.* 1985, Salwasser *et al.* 1993, Sample *et al.* 1993, Thorne-Miller and Catena 1991, Wallace 1994, Westman 1990a, and Wood 1994).

9. The concept of ecosystem health has been the subject of much consideration in recent decades (Callicott 1992, 1995, Christensen *et al.* 1996, Costanza 1992, Costanza *et al.* 1992, Ehrenfeld 1992, 1993, Hannon 1992, Hargrove 1992, Haskell *et al.* 1992, Karr 1990, 1991, 1992, Keddy *et al.* 1993, Norton 1991, Rapport 1989a,b, 1992, Ulanowicz 1992, Woodley *et al.* 1993), including the publication of a journal (EcoHealth).

10. CCAMLR, Canberra (1980) as reprinted in Wallace (1994).

11. See, for example, the U.S. Marine Mammal Commission's compendium of such agreements (Wallace 1994).

12. An essential part of the message of this book is the matter of learning from nature (Hobbs and Fowler 2008). This has been an approach to management advocated for centuries. Smith (1958) indicates that Taoism is basically ecological and "…seeks to be in tune with nature". Zeno (335–264 BC) maintained: "The purpose of life is to live in agreement with nature". Cicero (106–43 BC) added: "Those things are better which are perfected by nature than those which are finished by art". In the Bible, we are advised to "…ask now the beasts, and they shall teach thee; and the fowls of the air, and they shall teach thee; or speak to the earth and it shall teach thee; and the fishes of the sea shall declare unto thee" (Job 12:7–8). Later (about 75 AD), Plutarch advised that "The soul of man…is a portion or a copy of the soul of the Universe and is joined together on principles and in proportions corresponding to those which govern the Universe". Roger Bacon, in the thirteenth century, while advocating unrealistic control, nevertheless saw that we needed to conform to the laws of nature, suggesting: "Nature can only be mastered by obeying its laws". Various writers, philosophers, and thinkers have added their voices to the matter of learning from nature. Henry Miller is one: "The world is not to be put in order, the world is order. It is for us to put ourselves in unison with this order". Using nature as a teacher is carried forward in the writings of Henry Thoreau and John Muir who saw "…sensitive observation of nature as the source of wisdom" (Norton 1994). Aldo Leopold said: "A thing is right when it tends to preserve the integrity, stability and beauty of the biotic community. It is wrong

when it tends to do otherwise". Christensen *et al.* (1996) characterized Leopold's perspective as that in which wilderness provides a "base-datum of normality". Bateson (1979) indicated that we should follow the examples of nature. The use of nature as a source of guidance has been revisited numerous times (e.g., Chiras, 1992, Willis, 1995) and is embodied in the process of biomimicry (Benyus, 2002). The writings of people like Wendell Berry carry forward the philosophy of earlier luminaries (e.g., Berry, 2001, Berry and Wirzba, 2002).

13. A good match (consonance) between management question and informative pattern (and thus the science that reveals the pattern) involves common units, common logical typing, and common circumstances (Belgrano and Fowler 2008)—a one-to-one mapping or isomorphism.

14. Some aboriginal societies may be sustainable now. A book similar to this might be written by making comparisons with and among such societies (e.g., see Diamond 2004). To maintain a clear distinction and draw upon a much broader (and hopefully less contentious) field of knowledge, the comparative approach adopted for this book is restricted to interspecific comparisons. The challenges of achieving sustainability will be addressed again in Chapter 6 and in the Epilogue—challenges that may be beyond our capacity to meet.

15. This point will be amplified in later chapters. The basic idea is exemplified by the question: "What is an appropriate body temperature?" The answer is found in observations of body temperature, not a simulation model based on information or observations regarding physiological processes, genetics, environmental temperature, behavior, or chemistry (all related to body temperature) from studies conducted independently of body temperature.

Chapter 2

1. As introduced in Chapter 1, consonance between pattern and management question involves an isomorphism, or one-to-one mapping between the two. When consonant, they each have the same units, have the same logical type, fall in the same category, and involve the same circumstances (context).

2. A blue whale, weighing 136,000 kg, can be assigned the upper end of such a size range (recognizing that a few plants exceed this size). Thus, species in the lower 0.1% of the size range comprise organisms 136 kg or smaller. Based on the distribution of terrestrial mammals from May (1978, 1986), over 99% of terrestrial mammals are less than this size. Since insects (about 2/3 of all species; World Conservation Monitoring Centre 1992) and other invertebrates, as well as most other species (most other vertebrates, microbial species, most plants, etc.), are smaller than 136 kg, 99% is a conservative estimate of the portion of species within the first 0.1% of the size range.

3. Many examples of information regarding observed patterns in body size, with discussion of contributing factors and interpretations are found in the literature; some include consideration of life history characteristics such as generation time (e.g., Aarssen *et al.* 2006, Anderson 1977, Basset and Kitching 1991, Bonner 1968, Brown 1995, Brown *et al.* 1993, Damuth 1992, Dial and Marzluff 1988, 1989, Fleming 1973, Gaston and Blackburn 2000, Gaston and Lawton 1988a,b, Hutchinson and MacArthur 1959, Maurer *et al.* 1992, May 1988, Rosenzweig 1995, Sinclair 1996, Stanley 1973, Van Valen 1973a).

4. Such literature is abundant (e.g., Anderson 1977, Patterson 1984, Pimm 1982, Rosenzweig 1995, Strong *et al.* 1984, Sugihara *et al.* 1989, Warburton 1989, Yodzis 1984).

5. We most be mindful of the fact that virtually all materials consumed are recycled within the ecosystem where consumed. This is different from human harvest wherein much is removed from the ecosystem. Also involved here is selectivity—a topic involving a different pattern related to evolutionary impact of consumption, human or nonhuman.

6. Population density is the focus of numerous papers (e.g., Basset and Kitching 1991, Bock 1984a,b, 1987, Bock and Ricklefs 1983, Brown 1995, Collins and Glenn 1990, Damuth 1991, Gaston and Blackburn 2000, Gaston and Lawton 1988a, 1990a,b, Glazier 1986, Gotelli and Simberloff 1987, Hanski 1982, Hengeveld and Haeck 1981, Lawton 1990, Preston 1949, 1962, and the many papers that cite these in extending analysis and observations).

7. Data used to construct this figure are crude population estimates (from Nowak 1991, and Ridgway and Harrison 1981–1999) for species from approximately 20 kg to roughly 140 kg. The numbers are for species of mammals regardless of trophic level (includes both herbivores and carnivores) and habitats (e.g., marine and terrestrial). Midpoints were used where ranges were reported for both population and body size, mean masses were used to account for differences in size between the sexes. See Fowler and Perez (1999) for further detail.

8. White (1978) indicates that, among animals, true asexual reproduction accounts for only about 0.1% of the species. Other references to the preponderance of sexual reproduction are less quantitative (e.g., Mayr 1982, says "But sexual reproduction is by far the most predominant form") and Eldredge (1985), for example, says: "Indeed, strict asexuality (as opposed to some form of alteration of generations) is truly rare and, apparently,

always represents a secondary loss of sexuality". See also Buss (1988) and Maynard Smith (1989) and the review of Blackwelder and Garoian (1986). For very small species, it is possible that asexual species outnumber the sexual.

9. Blackwelder and Garoian (1986) review reproductive modes in respect to definition as well as incidence.

10. The distribution of species over metabolic rates per unit area could be presented similarly, using equations relating body size and metabolic rates but the distribution would be very similar to that of Figure 2.19 or 2.20 (offset by factors relating metabolism and density to body size).

11. A variety of ways to look at a stereogram to see the three dimensional image are presented in Horibuchi and Inoue (1994).

12. This is analogous to the "morphology" of a species as it would be seen in the frequency distributions among the individuals of that species. For example, if the weight, blood pressure, and pulse of a large representative sample of individual humans were plotted as in Figure 2.34, a three dimensional shape would emerge. With increasing weight (and concomitant age), blood pressure would rise and pulse would drop. A similar form would be observed for size, body temperature, and pulse. These relationships would give rise to form with variability about the underlying correlations. Within the scatter of points, as a pattern, the bulk of points would be concentrated among the smaller size ranges owing primarily to the age structure of our population.

13. These four domains are (1) continued work in conventional ecology (ecological mechanics), (2) evolutionary processes of natural selection at the individual level, (3) natural selection at the species level (selective extinction and speciation), and (4) the ways environmental factors are in involved in the first three. Listing such domains is not intended to imply that there are not interactions and synergistic effects/factors involved in any combination or subset.

14. The concept of emergence (Bateson 1979, Clayton 2004, Cowan *et al.* 1994, Emlen *et al.* 1998, Jørgensen *et al.* 1999, Lewin 1992, Morowitz 2002, Nagel 1979, Nielsen 2000, Prigogine 1978, Scheiner *et al.* 1993) is at the core of the integrative nature of patterns and will be treated repeatedly throughout this book. The infinite of Figure 1.2 is the complexity (or reality of Appendix 1.1) accounted for in its emergent patterns and emergence is one of the processes involved in reality.

15. Consider the example of the rate of predation by one species on another. This seemingly mechanical process is not unrelated to, or not without influence from, evolutionary changes that may lead to greater capture efficiency. The capacity for avoiding predation is also a matter of natural selection on the part of the prey species. Both, and the balance between them, influence the rate at which predation is realized and thus the location of a species on the relevant species frequency distribution.

Chapter 3

1. It might be argued that implementation of the U.S. Endangered Species Act, and research on population viability represent a huge effort and thus provide a counter example to this statement. These are largely mitigating efforts dealing with symptoms of deeper problems and do not manage to prevent the anthropogenic causes of reduced populations or endangered species in the spirit of Management Tenet 2 (Chapter 1).

2. These dynamics are parallel to some of the elements of population dynamics (specifically with the death and birth of individual organisms as contributing components). As with populations of individuals, "populations" of species in systems like ecosystems also experience movement (immigration and emigration).

3. Species frequency distributions are emergent from complexity (Fig. 1.4). The process of emergence (itself complex and involving reality: Appendix 1.1) is integrative. The integrative aspect of emergence is analogous to Bayesian statistical integration brought out later in this book and embodied in selective extinction and speciation described in this chapter (see also, Fowler 1999a,b, Fowler et al. 1999). This integrative process is also analogous to the formation of natural patterns as Nash equilibria (Fowler 2008, Nash 1950a,b), as pointed out at several points in this book.

4. At this point it needs to be emphasized that complexity includes hierarchical organization such that diversity among the components of any biological organization (e.g., species within an ecosystem) is only part of the picture. Species comprise individuals. But individuals are composed of organs, which involve cells, consisting of organic compounds, that are made up of elements. The matter of complexity in hierarchical organization goes the other way as well. The biosphere is made up of communities within ecosystems. These scales of structure are accompanied by scales of space and time.

5. Mentioning ecosystems without mention of other groups of species (such as those found in the biosphere) is not meant to exclude such groups—all groups of species evolve through selective extinction and speciation involving their genomes just as all groups of individuals do so through natural selection among individuals and the genes they carry. The effect of extinction on ecosystems depends on the kinds of species that suffer extinction (Hubbell 2001, Petchey et al. 2004, Solan et al. 2004).

6. This simple example is restricted to selective removal (death and extinction) so does not involve birth or speciation that are also part of the process of natural selection as a whole. A similar graph could be constructed wherein only selective speciation resulted in change. In reality, of course, both, in various mixes, are involved. Part of the oversimplification of this example involves the fact that for both processes of reproduction there can be a "backfilling" that will result in less change than depicted (recombination on the part of genetic dynamics among individuals, and possibly diversity dependent cladogenesis among species, for example). The point of the illustration is only to show tendencies that can result from both forms of selection.

7. Or "more making" (Eldredge 1985) through cladogenesis.

8. These exceptions can be either spatial or temporal. That is, some species within a particular set of species may evolve in a direction that is opposite the overall tendency just as some isolated segments within a lineage may show reversal of the overall trend over time.

9. See, for example, Charlesworth et al. (1982), Gould and Eldredge (1977), Maynard Smith, (1983), Slatkin (1983), Stanley (1975a,b, 1979, 1989), and Vrba (1980).

10. Often referred to as a rule, this is simply a statement saying that body size *tends* to increase more often than decrease over evolutionary time, both within and among lineages. For discussion of this phenomenon see Futuyma (1986a), and Gillman (2007) Newell (1949).

11. For further consideration of directional evolution, especially regarding specialization see Berenbaum (1996), Brasier (1988), Ferry-Graham et al. (2002), Futuyma and Moreno (1988), Gill (1989), Gould (1982a,b), McNamara (1990), Newell (1949), Stanley (1989), Travis and Mueller (1989), Valkenburgh et al. (2004), Vrba (1980), and Williams (1992).

12. Sometimes referred to as "Wright's rule" (Maynard Smith, 1983) there is ample consideration of what might be called randomness in evolutionary direction as exemplified by Charlesworth et al. (1982), Gould and Eldredge (1977), Maynard Smith, (1983), Slatkin (1983), Stanley (1975a, 1979, 1989), and Vrba (1980).

13. See also the references in endnote 11.

14. As will be apparent in many of the references of this chapter (especially Appendix 3.1), many people have contributed to thinking about the concept of selective extinction and speciation to provide descriptions of its fundamental elements. Listing specific references does injustice to those omitted. However, in the interest of providing a guide to the concept the following references should also be helpful: Bateson (1979), Brandon (1988), Damuth (1985), Eldredge (1985), Fowler and MacMahon

(1982), Ghiselin (1969), Jablonski (2007), Maynard Smith (1983), Okasha (2006), Reed (1981), Slatkin (1981), Vrba (1989), and Williams (1992).

15. This simply means that the interaction a species experiences with its environment in regard to selective extinction and speciation will depend not only on the environment but on the characteristics of the species. Property (attribute, characteristic) dependence is being emphasized more in this book than the effects of the environment. The latter would be very important in a book looking at the evolution of ecosystems from the point of view of comparative ecosystem studies that integrate selective extinction and speciation as influenced by environments of different kinds. See the sections later in this chapter dealing with the effects of the environment in selective processes at the species level.

16. This is obvious when we note the similarities of coral reef ecosystems regardless of their location, and similarity of desert ecosystems whereever we find them, compared to the differences between any coral reef and any desert ecosystem.

17. Another element in species-level dynamics, of course, is movement. Species move in and out of spatially defined species assemblages just as individuals move in and out of a population. This is an important element within the factors contributing to the formation of species-level patterns, but not one given its due consideration in this book. It is mentioned in this endnote to ensure that it is included in considering the complexity of factors involved. It is part of the category of ecological dynamics, and can be measured as a species-level attribute (mobility). At larger scales, of course, evolutionary changes that allow expansion or relocation of the geographic ranges of species also occur.

18. See endnote 3, Chapter 1.

19. This process (Fortey 1989, Raup 1984) has been described as "the evolution of one species into another" (Ehrlich and Wilson 1991). Another term "extinction by transformation" is used (but not accepted in principle) by Eldredge (1985). The definition of species one adopts may determine how many species will be dealt with in a species frequency distribution. But this is of little consequence to whether or not the dynamics of selective extinction and speciation apply as described and explored in this and the following chapters (Fowler and MacMahon 1982).

20. Any species-level characteristic can be involved in place of body size as used in the example here. Other characteristics were dealt with in Chapter 2. The dynamics of changing category are treated in the appendices of this chapter insofar as they involve extinction and speciation, but of course, they always include simple

ecological mechanics as well (e.g., changes in population size or geographic range).

21. Note that the word "similar" is used here because species at two points in a temporal lineage cannot be identical in all their traits but do resemble each other. Part of the difference between pseudo-extinction and terminal extinction is that the latter involves a more discrete end point marked by the death of the last individual, whereas taxonomists will argue over the definition of the point in time at which one species has evolved to become another through anagenesis.

22. The recognition of mass extinctions exemplifies knowledge about variability in extinction rates. Discussion and debate regarding mass extinctions abound in the literature; the topic of variability in extinction is central to many papers (e.g., Elliot 1986, Signor 1990). Recently variation in extinction rates has been associated with climate change (e.g., Benton and Twitchett 2003, Kiehl and Shields 2005, Ward 2007, Wignall 2001).

23. See Hull (1976, 1978, 1980) and others (Eldredge 1985, Salthe 1985, Williams 1992) for consideration of species as individual units (holons, see Wilber 1996) in this regard.

24. References treating the effects of the physical environment in speciation include Cracraft (1985a), Hallam (1989), Knoll (1989), Raup and Boyajian (1988), Signor (1990), and Stanley (1984).

25. The biotic components of extinction are embodied, in part, in the coevolutionary interactions (Futuyma and Slatkin 1983a) among species wherein one species experiences altered selective forces caused by a change in another. Literature on biotic sources of extinction is exemplified by Brooker *et al.* (2007), Futumya (1989), Hoffman (1989b), Maynard Smith (1976a,b, 1988, 1989), Mayr (1982), Miller (1956), Mitter and Brooks (1983), Rankin and López-Sepulcre (2005), Roughgarden (1983), Signor (1990), Simberloff (1983), Simberloff and Boecklen (1991), Stanley (1989), Stanley *et al.* (1983), Stenseth (1985, 1986, 1989), Stenseth, and Maynard Smith (1984), Vermeij (1983), Williams (1992), and Wilson (1980).

26. As indicated by Okasha (2006), the concept of "punctuated equilibria" has been a focal point of considerable palaeontological attention (e.g., Eldredge 1985, Eldredge and Gould 1972, Gould 1982a, Gould and Eldredge 1977, Hoffman 1982, 1983, Stanley 1989, Stanley *et al.* 1983).

27. We must keep in mind the hierarchical nature of this statement. There is variety *within* each distribution and there is variety *among* the sets of distributions. Each has its own mean, a mean that varies from set to set to contribute to higher level patterns of variability which may be correlated with environmental factors (e.g., as ecosystem characteristics that have their own patterns over space and time).

28. Wholes are always more than the sum of their parts—one aspect of emergence.

29. The confined variation within frequency distributions results in patterns that can be characterized as Nash equilibria (Nash 1950a,b, involving both individuals and species), but always one in which all factors are considered as part of the "game" being played in nature (Fig. 1.4). Thus, all ecological mechanics also play their role in producing nature's Nash equilibria (Fowler 2008).

30. The expense of sex at the individual level has been recognized for decades. The attraction of predation for many animals is only one of the disadvantages of sexual reproduction to individuals. Among other elements of the "widely discussed selective disadvantage of sex" (at the individual level of course, May and Anderson 1983, page 192) are the costs of parental investment in egg materials, dilution of genetic materials genotype mismatch with environment, and high mortality rate. There are advantages, to be sure, but as stated by Williams (1975) "The impossibility of sex being an immediate reproductive adaptation in higher organisms would seem to be as firmly established a conclusion as can be found in current evolutionary thought". "Ultimately, one would expect sexual reproduction to disappear from any species in which asexuality is an option" (Mooney 1993). In fact, a certain portion of species are known to lose their sexual reproductive mode through evolution over relevant time scales (Stebbins 1960). Sexual selection is recognized as a likely contributor to "evolutionary suicide" (Morrow and Fricke 2004, Morrow and Pitcher 2003). These difficulties give rise to an enigma when it comes to trying to explain the preponderance of sexual reproduction among all species. The dilemma is reflected in much of the volume of literature devoted to sexual reproduction and its evolution (e.g., Emerson 1960, Ghiselin 1974, Lewis 1987, Maynard Smith 1971, 1978a,b, 1988, Michod and Levin 1988, Nicholson 1960, Policansky 1987, 1987, Roe 1983, Schultz 1977, Shapiro 1987, Stebbins 1960, Werren, 1987, Williams 1975, Wilson 1975). It is within these considerations that the advantage at the species level emerges to be seen as a situation that falls into Combination 7 (or possibly, but not probably, 8) of Table 3.1.

31. It should be kept in mind that almost all species have gone extinct and that they possessed characteristics that arose through their evolution, primarily through natural selection among individuals. Today, we are seeing increasing awareness of the details of how this occurs as more attention is being paid in research to the concept of "evolutionary suicide" (see previous endnote).

32. There is confusion in the literature about the distinction between selective extinction and speciation

and group selection but the predominant focus of group selection is that described here (e.g., Arnold and Fristrup, 1982). For the variety of treatments, and consideration of the history of the concept of group selection see Darlington (1971), Eldredge (1985), Futuyma (1986a), Hagen (1992), Rosenzweig (1974), Roughgarden (1983, 1991), Salthe (1985), Travis and Mueller (1989), Van Valen (1971), Williams (1992), and Wilson (1983).

33. A number of references can be consulted regarding this point (e.g., Briggs 1988, Cambell and Clark 1981, Futuyma 1973, Howe 1977, Karr 1982a, Lovejoy *et al.* 1984, van der Maarel 1975, Martin and Klein 1984, Maynard Smith 1989, Owen-Smith 1988, Pimm 1980, Pimm and Gilpin 1989, Stanley 1984, Wilcox and Murphy 1985).

34. This literature also contains the variety of terms used to describe the dynamic of one or more extinctions resulting from earlier extinctions. They have been called ripple effects, cascading extinction (or cascading effects), cascade of extinctions (Owen-Smith 1988, Schindler 1989, Terborgh and Winter 1980), linked extinction (Gilbert 1980), and secondary extinctions (or secondary effects, Pimm and Gilpin 1989).

35. This is a common theme in palaeontological and ecological work (e.g., see Cracraft 1985a, Eldredge 1991, Hallam 1989, Knoll 1989, Parsons 1991a,b, Signor 1990).

36. Van Valen (1973b) described the process of species evolving in response to each other's evolution by quoting from L. Carroll's *"Through the Looking Glass"* ("Now here, you see, it takes all the running you can do, to keep in the same place."). Such processes are thus termed the Red Queen model of evolution (Maynard Smith 1988).

37. The development of the study of coevolutionary interactions and their role in speciation is far beyond comprehensive review in this book. The kinds of interactions thought to play roles in speciation as listed here are found in a number of papers (e.g., see Bakker 1983, Barrett 1983, Ehrlich and Mooney 1983, Futuyma 1983, 1989, Futuyma and Slatkin 1983a,b,c, Gilbert 1983, Janzen 1983, Janzen and Martin 1982, Maynard Smith 1976a, 1989, Miller 1956, Mitter and Brooks 1983, Signor 1990, Stenseth 1985, 1986, 1989, Stenseth and Maynard Smith 1984, Van Valen 1973b, Vermeij 1983).

38. This represents a form of ecosystem organization responding to the stressful (including extinction causing) effects of the physical environment (self-organization within context). It is an example of the kinds of patterns expected in complex systems resulting from the effects of stress as identified by Prigogine (Prigogine and Stengers 1984).

39. Grant presents useful references regarding the early history of elements of the idea of selective extinction and speciation. In addition to Darwin, Grant includes Gause

level in biological hierarchies. This is because the strength of effects of more inclusive systems is greater on species and individuals than the effects of individuals on species, ecosystems, or the biosphere (Allen and Starr 1982, Campbell 1974, O'Neill *et al.* 1986, Salthe 1985, Wilber 1996). The effects of many species on one are greater than those of one on many; one holon has less influence on its many counterparts than vise versa.

36. Recall that six of the eight combinations of forces from Table 3.1 involved opposing dynamic forces depicted graphically and quantitatively in Appendix 3.2. The implication here is that opposing selective forces contribute to the origin of conflict. It is natural. It occurs in the dynamics of ecological mechanics as well. The dynamic balance of carrying capacity, for example is a balance of a complex of opposing forces (e.g., species that supply food, pollinating services, dispersal etc. on the supportive side, and limits to food supply, pathogens, parasites, and predators on the negative side). And, it occurs between ecological mechanics and selective forces.

37. Many definitions of management are restricted to the control of human behavior since it is really the only aspect of management wherein control is real. The implementation of this aspect of management is beyond the scope of this book (but will be treated briefly in Chapter 6). It cannot be left to sociologists, economists, ethicists, legal experts, behaviorists, or any other science or scientist alone. It involves society, the people within it, and the changes they are willing to make as individuals in implementing species-level change. The changes individuals make, how they make them, why they make them, and when they make them are the subjects of study of fields such as those listed. But these fields of study are less the source of change than they are means of understanding it, documenting it, or characterizing it. Change of the type necessary to make a difference will depend on (but not be limited to) *realized* change within the individual human.

38. It assumes that, in nature, the inherent value of individuals, species, or ecosystems may easily be different from what we humans might decide. This means that one, more than another, may be more significant in ecosystem dynamics, for example. Extinction guarantees the loss of individual organisms and, vise versa, the death of all individuals guarantees the extinction of a species. The relative importance of each process, factor, and relationship is accounted for automatically, as is that of each component part of every system. It avoids human values that lead to problems because of inadequate consideration of complexity. It incorporates any values involved in sustainability.

39. In statistical work, Monte Carlo and Bayesian approaches to parameter estimation are similar to what happens in selective extinction and speciation. The abstract models of these analytical techniques cannot account for the total complexity of factors but the resulting natural patterns account for everything to which they are exposed—their full history and explanation. Species frequency distributions along any of their dimensions (attributes) are observations relevant to probability distributions. The most risk-averse traits are interpreted to be represented by the most abundant kinds of species. Cumulative risks contribute to preventing the accumulation of species in the tails of species frequency distributions. They are the results of nature's Monte Carlo experiments (part of which are the diffusion processes of the models of Chapter 3 that rely on elements of complexity as they are involved at all levels).

40. Here the probability of the data (one of the factors used in equations to derive Bayesian statistics) is not dependent on human sampling error. The trial-and-error processes are exposed to samples of actual reality (we assume that reality has a probability of 1.0) and not measures of it.

41. Care is necessary to be sure that fair comparisons are made. In this case, for example, we would want to avoid using sets of species that consume only body parts (e.g., suck blood, eat fins, or consume leaves). If our management question involves mortality of individuals in harvesting, the species set for comparison must also kill individuals from the resource species. The selective forces must be made comparable by measures of the same dynamics. We would want to make comparisons based on species of our body size just as we would want to compare pulse for individuals of the same body size.

42. One of the principles of desirable management often identified (see Appendix 4.3) is adaptability. In the face of uncertainty and ignorance, a trial-and-error approach helps learn what works and how systems respond to various influences. Adaptive management (Holling 1978, Walters 1986, 1992) is given prime importance as a way of dealing with many of the challenges faced in resource management (Christensen *et al.* 1996, Hilborn *et al.* 1995, Mangel *et al.* 1996). Although humans will have to be adaptive to take on this form of management, the point here is that other species adapt through their evolution. The adaptive aspect of selective extinction and speciation results in the persistence of species that not only solve problems faced in natural selection at the individual level, but also have properties that are adaptive at the species level (a solution to the games found in Nash equilibria, Nash 1950a,b). Existing species are the surviving examples of adaptive management already conducted in nature with results and information awaiting our use (Fowler 2008).

43. The information content of species frequency distributions is a bit like having information (Fowler 2008) about the hands held by other players in card games; we then play where the objective is to continue playing (Meeker 1997). The kinds of hands that consistently win form a pattern. By recognizing the pattern we can "stack the deck" in our favor. Being a stochastic process, however, we can never allow ourselves to believe that we have control. The game being played is a form of "infinite" game (Carse 1986) compared to the finite games that are exemplified more by the transitive approach. (See also Nash, 1950a,b, regarding the game theoretical aspect of nature's solutions to complex games as Nash equilibria wherein things that work do so because they work at all levels of biological complexity.)

44. The element of control again emerges because it is important to recognize the limitation of laws, regulations, and mandates established by governments as means of achieving control (especially over change). Laws and regulations are a form of control that is transitive if generated by government independent of public support, and more intransitive if it emerges through popular support. This emphasizes the necessity of including individual humans as critical elements in change because much of the hope that exists rests on the capacity for individual change.

45. In Chapter 5 (see Appendix 5.2 for the process of breaking down management questions into their components as well as the process for expanding questions), the protocol for management of the kind being defined here is presented.

Chapter 5

1. Here, "species-level" is used to refer to the most direct application of systemic management emphasized in this book—dealing with the human species. This is not meant to detract from the fact that it includes dealing with our species' interactions with other species, groups of species, ecosystems, and the biosphere. Nor is it meant to detract from the fact that systemic management also applies to individuals and their interactions with other individuals, species, and ecosystems. Current medical applications involve the importance of normal blood pressure, body temperature, and nutrient intake. Systemic management includes, however, personal matters as emphasized by the importance of control issues in the psychological realm (e.g., Beattie 1987, Whitfield 1987) as a matter of implementation of Management Tenet 2.

2. As infants we begin the process of self-definition, differentiation, and integration as we become aware of others of like kind (e.g., Wilber 1996). In the species-level analogy, self-awareness is clearly a growing part of what our species is experiencing (e.g., Darwin 1953, Tiger and Fox 1989). Comparisons of humans with other species (Fowler and Hobbs 2002, 2003, Chapter 6) show humans to be "out of bounds". These are simply measures and information of use in decision-making regarding our sustainability. Systemic management includes the objective of species-level applications parallel to individual-level applications, including cases where individuals within our species must change to achieve consistency.

3. This would be the combination ultimately achieved in extending "whole systems thinking" to fully include both the context of systems *and* the content of systems *and* their interactions. All extrinsic, and all intrinsic, factors, elements, processes, dynamics, interactions, and relationships would be represented as the factors that contribute to the formation of species-level patterns (Fig. 1.4). They represent both nature's Bayesian integration of complexity and nature's Nash equilibria (where things must simultaneously be for the good of both wholes and parts). They are the results of nature's adaptive management.

4. Adopting systemic management represents a clear replacement of the transitive aspects of current approaches. This is not the only major change, however, and probably will not be seen as the most significant of changes involved. As depicted in Figure 1.1, one current role of stakeholders is replaced by observed patterns.

5. There are no limits to full scale holism (systems are always within other systems) in seeing species-level patterns as emergent natural phenomena. The categories of ecological mechanics, and the evolutionary processes of natural selection at the levels of both individuals and species were listed in earlier chapters. If there are other factors, categories of processes, scales of time, space or organization, or other aspects of complexity to be brought to bear, they are not ignored. They are not ignored in understanding species frequency distributions as natural phenomena emerging from all such factors (Fig. 1.4).

6. See endnote 1, Chapter 4, regarding the fractal nature of this pattern from one level to the next—a pattern that connects (Bateson 1979). Consistency involves simultaneous resolution for the conflicts identified in Chapter 4 (i.e., nature's Nash equilibria are successes observed to account for all levels of biological organization and the associated risks, interconnections, and complexity).

7. Ripple effects (including all things variously referred to as unintended consequences, side effects, domino effects, or downstream effects) involve all relationships and interactions. For evolutionary interactions (including coevolutionary variation) ripple effects include change set in motion, that results in evolutionary change by other species that itself stimulates evolutionary change in still other species throughout any biotic system to include reciprocity so that there are consequences (feedback) to the species initiating the influence.

8. Our influence on ecosystems over time has proceeded from that of being largely mechanical, to include larger selective forces in coevolutionary dynamics, to now include the evolution of species sets through the extinction of species within them (e.g., those represented in ecosystems). Thus our influence has expanded, especially in its magnitude, in the various scales of time, space, and level of biological organization. Virtually all of the evolutionary change we are responsible for is irreversible.

9. Further precaution may be advisable in the form of overcompensation at first. For example, initial reductions in our population that are more than indicated by existing consonant species-level patterns (Fowler 2005, 2008) might have advantages over smaller reductions. This would better insure the recovery of ecosystems and their recovery might occur more rapidly so that any abnormal risks to future generations of humans (and ecosystems) are minimized earlier.

10. Here we are thinking of individuals as would be involved in the management embodied in medicine, participation in social systems, and family dynamics. Fully systemic management also addresses the matter of resources used to maintain human health, the advisability of purposeful extinction of other species to control disease, and the genetic impact on our species of saving human lives. The list of expanded questions is virtually boundless.

11. This is in parallel with management at the individual level wherein it is clear that the first order of business in repairing dysfunctional systems is mostly a matter of changing yourself rather than trying to change others— an alternative emphasized in most therapeutic work.

12. Things are interconnected, or at least much more interconnected than can ever be represented in any artificial model. This is a principle underlying all realms of management and is embodied in many of the sciences. It is a basic principle in ecology and is recognized in the field of physics (or science in general) in Bell's Theorem (Bell 1964). Species in ecosystems are interconnected in webs of coevolutionary interactions (Lederberg 1993, Thompson 2005).

13. Complexity that is intrinsic involves the "interiority" (Wilber 1996) of species and individuals. We are made up of organs, cells, and molecules (and all of the things that make up organs, cells, and molecules, including all interactions and processes). Species are made up of individuals and their components. These are taken into account as part of the ecological mechanics of processes

and interactions among units (holons) of such systems insofar as they contribute to the formation of species frequency distributions.

14. Recall the model for e^x of the last chapter. The complexity of reality (compared to a single number) involves all constants, variables, dimensions, relationships, hierarchies, scales, factors, and elements that cannot be represented in a model but which make up reality. Reality involves *dynamics* not captured in the static nature of the value of e^x, to increase the scope of the infinite in complexity. Reality includes the context within in which e^x occurs. Reality is the right hand side of the "equation" in Figure 1.4—the infinite of the universe of factors involved.

15. Thus, the reductionistic guiding information is not reductionistic in what is taken into account and is free of the human elements of decision-making (bias, politics, emotions, limitations). The management question addressed and pattern used to answer it *are* reductionistic—thus meaning that there is an infinite number of questions to be addressed in fully accounting for complexity. There is no end to the research that is needed.

16. At the individual level, the analogy is that of maintaining normal body temperature, and blood pressure, and body weight, and pulse, and blood sugar levels, and respiration rates, and variation in them—not just one or a few of them and not at *fixed* levels. Normal variation is also important. Furthermore, the addressing of such questions has to be expanded to ask whether management actions we take to maintain our health are sustainable—whether such measures (e.g., devoting resources, and the creation of waste, plus the long-term effects of medical measures) should be taken at all. Management at the species level must include CO_2 production, energy consumption, water consumption, range size, population density—the list goes on.

17. This "estimate" is based on an extrapolation of the relationship between population density and body size found in Peters (1983; his Fig. 10.3, page 169, where $D = 30W^{-0.98}$, D is measured in nos. km^{-2} and body mass is measured in kg). See also Appendix Figure 2.1.22 regarding total population size in relationship to generation time and, therefore, body size.

18. Note that we may see clear evidence of current problems in regard to factors such as total population biomass, biomass consumption, or energy consumption (as will be treated in Chapter 6) without relying on information regarding body size (e.g., Damuth 1987, Schmid et al. 2001). Precision regarding the extent of such problems will likely require information regarding any relationships between these species-level measures and body size.

19. The advice we would get in addressing the predation question this way would therefore be consistent with the guidance we would get when addressing another management question directly. The management question would be "How much should we reduce our harvest of [species x] if its population declines by [y] percent?" Direct measures of the reductions in predation rates observed in nature would provide the guiding information. Such reductions would have averages and variation, and would show empirically observable limits to such variation.

20. This is parallel to accepting as natural the increased respiration rates associated with physical exercise for individual humans. But, it opens the matter of management to other questions, in this case: "What portion of each day is optimally devoted to exercise with a respiration rate that is 20% higher than average? What is a sustainable total daily respiration rate? How many heart beats per day is most sustainable? In ecosystems, what is the most sustainable harvest during el Niño years? Is the advised species composition of catches different in el Niño years compared to other years?"

21. An important distinction here is that each application is a focused application while the guiding information holistically accounts for complexity with each of the infinite individual factors involved in an objective way.

22. Possible, here, is used in the sense that systemic management does not mean that we accept individuals with abnormal qualities any more than we accept abnormal qualities for our species. Thus, the transitive action of imposed starvation to solve the problem of overconsumption by our species is not an option while accepting starvation as a consequence of overpopulation is. Systemic forces will result in the reduction of the human population in what we will evaluate as abnormal events if we find no way to reduce it. Such a solution to the problem of overpopulation may be the only realistic option—it is what works for other species.

23. We can return here to the equation for e^x and point out that the contribution of each element of complexity in the model is added in proportion to its importance. Thus, in reality, the alignment of the planets is considered in proportion to the strength of its actual influence in determining the position and variation of species within species frequency distributions. The same holds for other elements of reality, exemplified by sun spots, the energy from the sun, tidal cycles, the strength of the carbon/oxygen bond in chemical compounds, pheromones, gravity, behavior, social facilitation, evolution, DNA, body size, strange attractors—again, the list goes on.

24. In other words, individual humans will have to contribute by falling in different regions of the spectra of natural variation for individuals until the needed change is accomplished. Time is a clearly important factor here as hurried change can place individuals completely outside such distributions to be abnormal—a situation to be avoided in systemic management (but accepted if imposed systemically).

25. What is being conveyed here is that, if the alignment of the planets has a relatively small influence compared to that of the population size of the predator in determining a predation rate observed in Figure 4.1, that difference is taken into account. If the alignment of the planets has a butterfly effect (Gleick 1987), that is taken into account in relation to its appropriate scale of time. Also taken into account are those cases wherein the collective influence of all butterfly effects are larger than any one obviously important factor (the latter being a case in which a factor is determined to be important by science, a process often restricted to showing [proving] effects are significant at the 0.95 confidence level, before they are accepted as real).

26. Systemic management is a matter of mimicry wherein the concept of biomimicry (e.g., Benyus 2002) is both refined and expanded to address questions of limits (Fowler 2003). Thus we can approach the issue of feeding humans *and* the issues of how many to feed, how much food should be extracted from ecosystems, and how much should be extracted from the biosphere. It therefore becomes *self-limiting* so that any one form of biomimicry does not cause problems at other levels of biological organization. There is consistency.

27. That is, we cannot know all of the detail involved in the complexity of such systems, and any advice generated without consonant information produced by a particular field of science, or set of such fields of sciences, represents a reductionistic error in ignoring what would be considered by all forms of the sciences that would be focused on what we do not know now, if such fields of sciences existed. Advice stemming from what the various fields of sciences can measure as consonant information is different in that the measurements are of things that themselves reflect complexity, account for complexity, and result from complexity. This same complexity is reflected in nonconsonant information, but has to be converted—an alchemical process.

28. This limitation of science has been referred to as the Humpty Dumpty phenomenon (Fowler 2003, Fowler and Hobbs 2002, Nixon and Kremer 1977); scientists cannot combine their concepts and information to fully replicate reality (Bateson 1979). Each represents the study of parts of reality and each is finite as is true of any human construct (e.g., mathematics is also finite in this regard: Gödel's Theorem, Gödel 1931, Makous 2000).

29. Whole systems thinking (e.g., Wilber 1996) goes beyond the approaches of Brown (1995) and Rosenzweig (1995). These latter examples serve as notable steps toward "macroecology" as a synthesis of biological thought. But whole systems thinking is not restricted to biology, even if much of complex systems science can be traced to roots in biological (especially ecosystem-level) thought. The complex systems we often think of are themselves within larger contexts, larger systems. Inclusiveness is elusive; temporal scales are not to be ignored. Whole systems thinking is not particularly familiar to science as practiced in the last century or so. It is not so foreign, however, to religions and philosophies that grapple with "the Great One (or Oneness)", the "unspeakable infinite" (note the infinite of Fig. 1.4), Supreme Being, etc. It is not clear how holistic it is possible, or necessary, to think. To the degree that there is (or we can find) a common denominator in wholeness or synthesis (infinite) among the sciences, religion, and philosophy, there is hope of understanding empirical patterns as emergent sources of guidance (and maybe common ground between science and religion).

30. See Chapter 3, endnote 2 of Chapter 1, and Appendix 4.4 (also involving Nash equilibria, Nash 1950a,b; and adaptive management, Holling 1978, Walters 1986, 1992).

31. We are reminded of experiences exemplified by proving that smoking tobacco and using DDT are harmful after a history of resisting the possibility. The law of unintended consequences dictates that we accept that the reality of effects of things that we do that have not yet been proven scientifically to be real. This is integral to accepting the combination of complexity and interconnectedness.

32. This emphasizes the importance of the combination in what is studied by science (reality), rather than the combination of the products of science, when it comes to management.

33. A variety of authors have considered the matter of emergence, its relevance to various levels of biological organization, and its role in natural selection—also at various levels. Distinctions are sometimes drawn between aggregate and emergent properties. In the sense of Figure 1.4, this book remains open to the option that everything is emergent and that all components of complexity can be involved (Morowitz 2002). For various views and considerations consult Allen and Starr (1982), Bateson (1972, 1979), Brown (1995), El-Hani and Emmeche (2000), Eldredge (1985, 1989), Emlen *et al.* (1998), Fralish *et al.* (1993), Jørgensen *et al.* (1999), Kauffman (1993), Levinton (1988), Lewin (1992), Mayr (1982), Morse (1993), Nielsen and Müller (2000), Nixon and Kremer (1977), Pagel *et al.* (1991a,b), Salt (1979), Salthe (1985), Scheiner *et al.* (1993), Smith (1977), Stanley (1989), Vrba (1984), Whittaker (1975), Wilber (1996) and Williams (1992).

34. This information can apply at the various levels of biological organization. We already know about such work in medicine through research on individuals. But, information is also embodied in measures of species in species-level patterns to include an integration of the genetic information emergent from the Bayesian form of trial-and-error processes of selective extinction and speciation (as described in Chapter 3, Appendix 4.4). Or it can be integrated information that emerges as ecosystem-level descriptions. This is information *about* ecosystems. Measures of ecosystems are the result, and the dynamics of ecosystems as whole units (i.e., changes in the measures of central tendency for species-level patterns over time) may be monitored and described so that we can determine the normal range of natural variation for such systems.

35. The unknowable of the future is included only insofar as the past can be used to predict the future—as exemplified by seasonal cycles in both abiotic and biotic systems. This emphasizes the need to use correlative information regarding such things as weather to characterize frequency distributions under various conditions of the physical environment.

36. These include the work showing relationships between body size and other traits such as age at first reproduction, rates of increase, generation time, density, population variation, and geographic range (Chapter 2, Peters 1983, Schmidt-Nielsen 1984, Sinclair 1996). Food web analysis has resulted in demonstrating other patterns regarding trophic level, species numbers, connectivity. Continued ecosystem ecology can be expected to see more relationships associated with abiotic factors in geographic heterogeneity (e.g., latitudinal gradients in species diversity, Rosenzweig 1995).

37. It also, of course, involves cellular biology, medicine, physiology, chemistry, physics, molecular biology, atomic physics, nuclear physics, and all of the other sciences that deal with the facts regarding what we are made of, our origins, and nature.

38. Clear indications of the direction we need to take are presented in this book. Included are initial indications of the magnitude of needed change. However, specific goals or exact/fixed endpoints are not presented. It is clear that reducing CO_2 production, energy/resource consumption, and population size/density are among the elements of primary importance in starting. The immensity of needed change in each case is approximated. Where we want to end up more exactly should be (and can be no other than) a decision based on further consideration of complexity through both refinement and expansion of various related management questions. Also included, would be factors such as the standard of living wanted by society (our species) as brought to the asking of management questions—but only as a factor to consider as a matter of fine-tuning where we might fall within the normal ranges of natural variation (e.g., having even a smaller population than might otherwise be optimal in order to keep energy consumption, CO_2 production, biomass harvesting, and other related factors within their normal ranges of natural variation).

39. In addition to these factors, is the observation that the maximization of probabilities within a probability distribution or maximization of biodiversity within a biodiversity curve (e.g., Fig. 5.3, Fowler 2008) does not involve a sharp peak. We may be lucky enough to have a great deal of relevant data for a highly refined management question. However, the peak of a probability distribution or biodiversity curve is a single number surrounded by a range of other numbers with nearly equivalent probabilities or biodiversities. Measured in log scales, there is a great deal of latitude for options (although quite small in comparison to the extent of human abnormality). This is exactly as it is for body size, blood pressure, and body temperature in regard to individuals. We would be hard pressed to maintain our body temperature at exactly one value for what we consider equivalent circumstances.

40. We are reminded here that another contribution will be made in the changes *individuals* can make. In North America our per capita consumption of energy is about 93 times larger than that of other comparable species (18-fold, worldwide; Chapter 6). However, as will be made more clear in Chapter 6, reducing our per capita consumption to 1% or 2% of current levels is only a tiny fraction of the change required for the human species to consume energy at a normal rate. In spite of what appears as a major per capita component, however, our per capita consumption accounts for very little of the discrepancy between the total consumption by humans and the mean of that by other species (Fowler 2008; some of which is attributable to their underpopulated condition owing to abnormal human influence). This emphasizes the importance of the population factor (over 99% of the difference between humans and other species). The per capita consumption would need to be cut in half 4.16 times while the total would need to be cut in half 10.9 times, or 6.73 halvings through population reduction to match the mean among other species.

41. At the species level, we might choose to act on population size, initially, with more emphasis than on geographic range. But such guesswork is dangerous because the numbers of diseases we are exposed to in our broad geographic distribution, the degree to which we transport

species (increasing their mobility), the numbers of ecosystems we connect via our broad distribution, and other negative effects could very well be more important than taken into account by such a decision. Precaution dictates that we do all we can to achieve a normal position within species-level patterns for every way we find ourselves to be abnormal in all the ways we can find to measure ourselves as a species.

For individuals some characteristics, such as hair length, may not be as important as others such as body temperature. This has to be the case for species as well. Until we can sort out which species-level characteristics are of lesser consequence it is precautionary to adhere to as many as possible.

42. Benchmarking is a process used in business management wherein other businesses are mimicked where the other businesses are interpreted to be successes (Bogan and English 1994, Boxwell 1994, Camp 1995, Spendolini 1992).

43. Part of the precautionary measures necessary here involves being open to the possibility that other predators on walleye pollock will increase in abundance to cause walleye pollock to decline in abundance, resulting in a decline in predation by other predators. Under such circumstances, the total take of walleye pollock based on the species frequency distribution we have today might be an overestimate of sustainable harvest levels.

44. The mean consumption rates from walleye pollock, herring, hake, and mackerel by marine mammals (see Fig. 2.6 and Appendix Fig. 2.1.5) are 0.45%, 0.18%, 2.5%, and 0.18%, respectively. With enough comparable sets of data, any patterns in relation to life history strategy, body size, and environmental factors would be very helpful in guiding management. Although we are dealing with only a tiny sample in this case, it is of note that the harvest rate of 13% for walleye pollock in recent years is well beyond the range of variation for the mean consumption rates taken from these species by marine mammals.

45. Thus, selective extinction will often act to remove species with characteristics emergent at other levels. Extinction risks would be higher among species with chaotic population dynamics than species with less variable populations, for example. Those that go extinct would do so even though such dynamics may be the result of circumstances involving ecological mechanics and evolutionary processes through natural selection at levels hierarchically below the species. Extinction places constraint on the potentials that are exploited (explored) in the diffusion-like processes of natural selection at other levels. In this sense, selective extinction and speciation supersedes the mechanical and lower level evolutionary processes, even depending on them for a supply of species over which to exert constraint. Evolutionary (e.g., through natural selection among individuals) processes that produce extinction-prone species cannot overpower the risk. Ecological mechanics that result in extinction cannot override extinction. On the other hand, extinction can weed out such species so that the species left behind may be less prone to such risk, thus molding the set of species in a way that is more powerful than the other processes when they oppose each other. In other words, tendencies toward producing extinction proneness (Categories 5 and 6 in Table 3.1) may be reduced, but more through the extinction process than through processes at lower levels.

46. There must be a pair-wise correspondence between management question and the matching research question. There must be a similar correspondence between pattern revealed through research and the objective revealed. This correspondence is consonance in which there is an isomorphism that precludes stakeholder translation conversion typical of conventional management (top row, Fig. 1.1). Consonance involves identical units, the same logical types (Bateson 1972), and the same circumstances. There must be a one-to one mapping between both questions, and both the pattern and management objective. The refinement of management questions must be reflected in parallel refinement of the science question. Refinement makes the question(s) open to correlative information—uses correlative information when it is known to occur.

47. The issue here is parallel to the matter of not asking scientists (or medical doctors) to use information about metabolic chemistry, cellular structure, and anatomy to decide what we should use as the proper body temperature for humans. Instead we ask them to collect and interpret data dealing directly with the mean, variance, and normal limits to natural variation in body temperature for humans. The general question might be refined in parallel to that for the harvest of walleye pollock by asking "What is the appropriate body temperature for a 5 year-old boy, playing basketball, at an ambient temperature of 33°C?" Gender, age, weather, and activity now come into play, but in all cases where we find a body temperature of 42°C, action is taken to solve what is recognized as a problem.

48. Individuals practice systemic management in learning from other individuals, groups learn from other groups (e.g., married couples learn from other married couples, teacher/class combinations learn from other teacher/class combinations, communities learn from other communities, cultures learn from other cultures, nations learn from other nations), businesses from other businesses (in the benchmarking process), and our

species is in the position of learning from other species. This is all a matter of seeing the collaborative learning process (Goldberger et al. 1996) as one that crosses hierarchical/holarchical boundaries to apply at all levels of organization.

Chapter 6

1. The word "failure" is used advisedly. It is important to distinguish between 1) the value of models as tools in research and 2) what we now see as the erroneous belief (what might be called *fatally* erroneous belief) that it is safe to use models with output that is not consonant with the management questions. The empirical failure of such approaches is becoming widely documented and recognized (Frank and Leggett 1994, Larkin 1977, Ludwig et al. 1993, Magnuson 1986, Malone 1995, McIntosh 1985, Moss 1989, Nelson and Soulé 1986, Peters 1991, Pilkey and Pilkey-Jarvis 2007, Rosenberg et al. 1993, T. Smith 1994, Walters 1986). What look like successes may simply be cases wherein problems have yet to be manifested or discovered. Type II errors in detecting problems (Mangel et al. 1996, Peterman 1990, Peterman and M'Gonigle 1992, Peters 1991) prevent a full appreciation of their prevalence. This emphasizes the importance of precautionary measures in practical application of limited knowledge (Holt and Talbot 1978, Mangel et al. 1996). A major point of systemic management is to use empirical information, rather than simulation models, to be as precautionary as possible. As will be developed in later sections, a major problem introduced earlier in this book is that of category confusion in which models and management question are not consonant (they involve different units and are not isomorphic).

2. The environment also elicits phenotypic expression. In other words, ecological mechanics (various ecological and physical processes, including such things as sunspot cycles, regime shifts in the climate, decadal oscillations in the weather/climate, predator/prey dynamics, and top-down and bottom-up trophic effects) are also involved in the processes that result in species-level patterns as information on the limits to natural variation. They are part of the complete suite of factors that make patterns what they are (Fig. 1.4).

3. Further evolution is likely to alter the position of individual species within species frequency distributions. However, constraints, such as those of selective extinction and speciation, are likely to prevent the shapes of such distributions from changing markedly. These changes (shapes of distributions such as Fig. 2.34), involve ecosystem evolution, one of the factors that cannot be discounted (Chapter 3). They are automatically accounted for in systemic management to the extent that evolutionary changes in ecosystems are involved in the full suite of factors at play in the origins of species-level patterns (Fig. 1.4).

4. This is in contrast to using the mean of the log transformed version of these data (geometric mean). Both the distributions themselves, and the central tendencies of such distributions (and their relative utility) are among the details of the application of systemic management that will be subject to fruitful debate and further research. For example, the point at which humans could be located within frequency distributions to maximize biodiversity or the information content of such distributions would be a consideration (see Fig. 5.3, Fowler 2008).

5. The larger take in biomass, of course, is taken because of the monetary value of reproductive aged fish which are taken in large numbers—a conventional management decision. Is it a sustainable mode of harvest? We encounter new questions (dimensions) that need addressing. What is the most sustainable age (or reproductive) composition of a harvest? If we remain fixed in taking only adults, what is the most sustainable harvest when confined to taking adult aged individuals? We would need empirical distributions among predatory species showing the mean age (or reproductive value) of consumed fish to address the first question. Observed patterns for predation rates on adults (fish of the age/size taken in commercial harvests) would be needed for the second question. Care must be exercised here, as in all cases. Determining what is ultimately sustainable will require information on the limits to natural variation that is representative or typical of systems free of abnormal human influence. Also, we will be best served by data for species otherwise similar to humans in regard to things we cannot change easily (e.g., our body size).

6. W. Overholtz supplied information (personal communication, 1/24/97) regarding the approximate mean biomass levels for each of the four species and information on takes by commercial fisheries used to determine the fraction of the total catches that applied to each species consistent with the total from Overholtz et al. (1991)

7. The concept of consonance was developed in some detail in Chapter 5. Basically it means finding informative patterns that are expressed in the same units and have the same logical type (have the same category), as defined by the question. Finding consonance between question and pattern results in an isomorphic match. See Fowler (2003) and Fowler and Smith (2004) for further details.

8. Note here, that for one species, an ecosystem could be defined as the geographic area (plus all biota and environment) made up of the union of all the geographic ranges for the set of species with geographic ranges overlapping that species' geographic range. There are many other alternatives for defining an area to be called an ecosystem, even the arbitrary nature of political boundaries for nations. In systemic management, each would be treated consistently with all others.

9. The departure of humans from the mean is based on comparisons with other mammals without regard to body size. Including species differing from humans in body size is an option here, in part, because of the lack of a relationship between body size and energy consumption per unit area (see Damuth 1987, 2007) thus making energy consumption something that can be compared to a large variety of other species.

10. For the energy used by humans in addition to metabolic energy, the 1992 values reported for the world consumption of energy (World Almanac 1993) was divided by the total human population (5.7 billion) to achieve a per capita use of energy. This includes energy as fuel provided by ecosystems (e.g., wood, beasts of burden) thus having two kinds of impacts: the acquiring of fuel products and the impact of its later use.

11. In other words (as will be seen in later sections of this chapter), if the standard of reference for population density were measured in terms of its impact on biotic environments (rather than simple population density), the density that is most risk averse **and** maintains a standard of living comparable to today's human society might be one to two orders of magnitude less than would be indicated based on the patterns in density related to body mass (e.g., Fig. 2.31; or sustainable density as will be seen in Table 6.2). The actual reduction needed could be much more owing to nonlinearity. This enters the arena of tradeoffs between per capita (individual) and population (species) issues that involve hierarchical considerations to be amplified later in the chapter.

12. Sixty percent of the primary production distributed among 20 million species is to be compared to 40% for one, humans. This represents a difference of seven orders of magnitude $[(0.4/1)/(0.6/2 \times 10^7) = 1.33 \times 10^7]$ between the mean for other species and the use of primary production by humans (and human dominated domestic species). With repeated cycling through ecosystems and degradation to heat, some of this difference will be lost. Even if the difference were an order of magnitude (it is a difference of less than a factor of 1.2 if 10% of the energy is passed between trophic levels) the remaining difference of six orders of magnitude means that humans are still monopolizing energy at a level where our energy use is a million times greater than that of nonhuman species.

13. From this point on, the issue of overpopulation is referred to as a problem in that our population is abnormally large (Fowler 2005 and the results of this chapter)—a pathological situation. Usage of this term can be seen as a matter of prejudice. This prejudice is justified on the basis of the comparisons between humans and other species in this chapter rather than on opinion. Calling overpopulation a problem is not meant to indicate or imply that there are not advantages (in human value systems) to our current population level in the complexity of such issues. The argument is that systemic consideration would indicate that, all risks and all benefits considered (Fig. 1.4), the risks outweigh all the benefits so as to include risks (and benefits) to other species, ecosystems, and the biosphere.

14. That is, it is no more possible to make a complete list of all factors involved in the complexity of reality, than it is to make models of it. Even with partial lists of the factors we know about, there is no means of assigning relative importance to each one. There is no methodology of human design that can replace empirical observation to account for reality in the way depicted in Figure 1.4.

15. A decline to 0.001 (0.1% or 6 million) of the current population would require approximately ten halvings, or about 300 years if it were reduced by half every 30 years (about the rate at which it was doubling in the last decades of the 20th century).

16. Furthermore, the alternative for reducing the population by increasing mortality through direct action will always be rejected when people consider how to reduce the human population. Any species without a genetically based opposition to mortality is unlikely to have survived. Natural selection at the individual level can be expected, more often than not, to produce opposition to death especially among age classes prior to achieving maximal reproductive value.

17. A precipitous depletion of fossil fuels would lead to identification of many of the species level risks currently and temporarily avoided, an experiment that may be worth trying to avoid but with a risk that may be out of our control.

18. Recently a great deal of effort and thought has gone to the issue of ecosystem health. During its short history, the International Society for Ecosystem Health specifically focused on this issue as do current journals such as EcoHealth and the consortium it represents. From the perspective of ecosystems as evaluated here, many or most ecosystems are ill and the disease is human overpopulation (magnified by, and, in part, caused by energetic/technological means, per capita consumption/production, and the other departures from normal variation for other species-level and individual-level dimensions). The problem of overpopulation is often referred to as an ecosystem-level cancer (see the references in

Hern 1993). However, the analogy fails in the sense that for individuals, if a cancer is not removed (or otherwise treated), death is a common outcome, whereas for an ecosystem to die, all populations of all relevant species would have to drop to zero (this can include humans, Boulter 2002). The ecosystem disease represented by humans is more likely only to result in losses of certain species and gains in others until humans count among those that are removed or reduced, much like individuals deal with some diseases through their immune systems. Following such ecosystem changes, the remaining species are left for evolutionary change to re-establish a more sustainable form of ecosystem adapted to the respective physical habitat and perhaps more resistant to the invasion and effects of species as disruptive as humans have been recently. In any case, it is clear that ecosystems are ill and we are the disease.

19. The disruptive effect of the human population has also been considered analogous to an explosion or "population bomb" in attempts to communicate the problem to an audience beyond scientists (Ehrlich 1968, and Ehrlich and Ehrlich 1990). The analogy is apt in that the volume of gases produced in an explosion is three to four orders of magnitude greater than the explosive material in the initial devise (Van Nostrands Scientific Encyclopedia 1958). Part of the explosive force is created by the heat of the resulting gases as an amplifying factor, much as energy, technology, and affluence amplify the effects of our population. The effect of an explosion can be measured in terms of rock moved or broken, for example, much as can the impact of humans be measured in terms of rates of extinction caused (8–12 orders of magnitude beyond the effects of the mean for other species, Appendix 6.6).

20. This is one of the examples that may apply as group selection among species. Those groups of species that contain species capable of removing disruptive species may be groups (e.g., the set represented in an ecosystem) that contribute more species to later "generations" of ecosystems. In this sense, the matter of overpopulation is probably beneficial to ecosystems in the long run as it will help produce ecosystems that reduce the risk of any similar species emerging to threaten the destruction of all of life. Part of what contributes to evolutionary change (now at the level of groups of species, including those defined by ecosystems) is the stress of factors involved in selective forces. Just as the immune systems of individuals can be stimulated toward greater efficacy by exposure to pathogens, so ecosystems are probably effected by exposure to overpopulation in their history.

21. In order to do our best at dealing with complexity, we must be very careful here to resist the temptation to take a transitive approach in "improving" the reality (change species-level patterns by altering nonhuman species and ecosystems) upon which they are based. The intransitive of systemic management is to manage ourselves so as to relieve systems of human abnormality so that they can organically respond, and, after such response, to reveal more accurately what is sustainable. The science involved always involves monitoring to reveal what is normal, follow change, and account for correlative factors underlying species-level patterns consonant with management questions.

22. Energy consumption per unit area was estimated by multiplying ingestion rates found using the equation from Peters (1983; $I = 10.7W^{0.70}$ where I is the ingestion rate per individual in Watts, and W is body mass, kg) by the densities reported in Damuth (1987) for the herbivore mammals of his sample, using the corresponding body mass (?). The estimated energy ingested per unit area by humans was based on assuming a density of 250 per square kilometer by assuming that the human population occupies a geographic range within the normal range of natural variation among species otherwise similar to humans (Appendix 6.4).

Chapter 7

1. This happens through breaking down every management question to its components (Chapter 5). Ultimately, these deal with the specific aspects of implementation—fractal application. Guidance is found in the detail where action can be taken, empirical information, but always using the processes outlined in this book for the overarching issues. Laws, regulations, and decisions about daily life would all be based on past experience and knowledge of what works. The importance of role models and mentoring has long been recognized in business, leadership, family, and social dynamics. Decisions and actions are based on knowing the difference between failed and successful approaches, again a trial-and-error approach, but used only insofar as they work to achieve goals established in consideration of higher-level questions.

2. The psychological aspect of change involves realizing that we are basing our management decisions on what are basically illusions (advice for management and policy based on the magical thinking involved in conversion of partial, nonconsonant/dissonant, information—top row Fig. 1.1). The illusion that such an approach is acceptable may be the key illusion to overcome and involves very personal kinds of change. Until we experience this as real cognitive dissonance, this critical step is unlikely to be taken—especially at the species level. But there is more. The psychological aspect of change also involves

knowing that emotions are to be used as motivation for asking the right questions; they are not to be, themselves, among the things converted directly to action.

3. The "course" has been identified in numerous analogies. The trajectory of our species has been likened to the Titanic, headed toward an iceberg, where our actions in conventional management amount to no more than the matter of rearranging the deck chairs (e.g., Parenteau 1998). We are seen as fiddling while Rome is burning, or shoveling fuel to a runaway train (Czech 2000, 2006). We seem to be the proverbial "bull in a china shop"—a problem we attempt to solve by moving a few bits of the china rather than putting the bull where it belongs (with science that studies the breaking of the china as science that is provided with major funding to help decide which china should be moved first).

4. The choice we face as a species is much like the choice faced by individuals who are told they are overweight, or that smoking is bad for their health. Many individuals in such situations make no change in their lives—choosing their life as they live it over what it could be otherwise. Likewise, we can continue, as a species, ignoring signs that there are risks we might wish to avoid. Management is a matter of choice—individual, social, political, national, and international. As a species, we face the message that we are a malignancy—a message to be accepted by us as individuals and taken to the species level.

5. It would not be unnatural, for example, to experience global warming as a self-exaggerating process in CO_2 contributed to the atmosphere by volcanoes. We know that the redistribution of mass on the Earth's surface results in isostatic readjustments (e.g., Whitehouse *et al.* 2007). If the trillions of tons of water from melting glaciers press tectonic plates into accelerated subduction, we should not be surprised to see increased earthquake and volcanic activity, the latter of which is known to add CO_2 to the atmosphere. As always, however, things are never simple: volcanoes also produce other gases that counteract the greenhouse effect of CO_2.

6. Recall Table 3.1 and the cases where evolution leads species to extinction—evolutionary suicide or fatal flaws that emerge through natural selection among genes and individuals (Potter 1990). The hope here lies in our being able to exercise existing intellect to our advantage. However, it is also possible that evolution has not yet completed the process of preparing us for the integrative thinking necessary to avoid causing our own demise. It is possible, for example, that integrative thinking is dependent on coordinated function of the two hemispheres of the human brain and that evolution has not yet brought us to the point of having the necessary intellect to do what is good for all living systems so as to include those of which we are a part and upon which we depend.

7. There is even a specialized journal devoted to consideration of the concept and its importance (*Emergence* is a publication of the Institute for the Study of Coherence and Emergence published by Lawrence Erlbaum Associates. Inc., Publishers, Mahwah, New Jersey; ISSN 1579–3250).

8. This is important to not abandon thinking. Thinking about thinking, addressing the epistemological aspects of what we do, is critical. Thinking is crucial to the conduct of good science, and the choice of science to provide information consonant with management questions. Thinking is a serious problem, however, as a way to combine and convert nonconsonant information to consonant information. We continue to rush into trouble by training more people to practice such alchemy (often seen as an important function of our educational system in conventional terms). The alternative is to train specialists to observe (see rather than think) so that guiding information is produced by good science—information consonant with good management questions.

9. This would follow the protocol developed in Chapter 5, exemplified by science to collect data on patterns for population size when the question involves establishing a sustainable population for humans. Succeeding analyses would involve accounting for patterns involving body size and other correlative patterns of relevance to humans and environmental factors. However (see endnote 1), the data needed are those *directly* defined by (consonant with) the management question (i.e., not component, tangential, merely relevant, or connected questions). The factors that are not consonant with one question would be consonant with another—there is a reciprocity.

10. The contributions of medicine to our dependency on other species, artificial chemicals, and changes in our genetic makeup (the evolutionary effects of medicine on our species) are also part of what must be covered in course work.

11. Between 1993 and 1994, books were listed in *Science* magazine (after having been sent to them by publishers) at the rate of about one every 45 minutes of the 40 hour work week. Scientific papers were published at the rate of one every 18 seconds in the early 1980s (Bartholomew 1986). Books with topics related to ecology were logged into the University of Washington library system at the rate of over 400 per year between 1994 and 1997. This material, gleaned from measurements and observations, is now to be complemented by the encyclopedic volume of information contained in each species' genome (e.g., house mouse with equivalent of 15 editions of *Encyclopedia Britannica*, Wilson 1985) and the unfathomable combination of such information in the form of species-level patterns (Fig. 5.4).

12. Consistency in objectives is achieved in systemic management (Management Tenet 4). One of the consequences would be markedly reducing conflict among special interest groups, political factions, religions, and other groups founded on various anthropogenic, and very often anthropocentric, belief systems (Hobbs and Fowler 2008). Adopting systemic management would not only put environmental organizations on a common foundation, with common goals, but would also greatly alleviate conflicts between environmental organizations and governments, between government agencies, between environmental organizations and business/industry, and between government and economically driven institutions. There might be divergence of opinion regarding what questions to ask in finding information relevant to implementation, but the objectives would be held as common objectives, even among nations, religions, and races.

13. Religion has strayed by adhering to beliefs (realities) that have little or no counterpart in reality and by losing a great deal of reverence for the nonhuman—even the holistic which distinguishes religion from science. Science is reductionistic and seems to be regaining lost commonality with religion in understanding that everything is emergent in a way that accounts for complexity—including the nonhuman elements of reality and reciprocity of relationships (Fig. 1.4).

14. Does the infinite of ultimate reality (Fig. 1.4) have anything in common with religious belief systems (e.g., ultimate reality being God)? Can scientists and members of religious organizations realize common ground in this regard? What role can religion play in social change?

15. What balance between standard of living and population size do we want for ourselves? How do we insure an acceptable standard of living for both other species and future generations of our species? Can our cultures and society find common ground in reality to move forward? Can we see a biotic democracy in the information presented to us by other species? What social systems have endured long enough to provide evidence that they work (are sustainable)?

16. Can there be unanimity among individuals and social factions to move stakeholders from decision-making to question-asking? How, besides taking a systemic approach, are conflicts between differing views to be resolved in ways that provide an objective basis for decision-making? What are the political systems that lead to change and adaptive strategies that will work?

17. Is it right to fight for the freedom of giving birth and providing food for everyone if it contributes to the risk of extinction, particularly human extinction? What are the rights of other species and the individuals within each?

18. Under systemic management, laws such as the Endangered Species Act in the United States would not be necessary as we would be dealing with all of the anthropogenic contributions to endangerment we can possibly imagine. Efforts to save endangered species are well intentioned and may even be a way to help get some species through the period of time during which we are solving anthropogenic problems that contribute to their endangerment. However, attempts to save other species count as mitigating actions that cause other problems and have their own unintended consequences. If we were a species free of pathological attributes for everything we can measure now (e.g., a population of two to ten million people, appropriating one ten millionth of the Earth's primary production, and occupying a geographic range of ten million square kilometers) there would very likely be little concern about endangered species.

Can we make our laws consistent with the laws of nature?

If we were managing systemically we would be playing the game as an infinite game (of which there is only one, Carse 1986) and there would be no need for laws to protect wilderness because people would understand that the game is to be played so as to continue the game, not to contribute to risks in which the game can no longer be played (Clark 1989, Eldredge 1991).

We already know how to make laws that help achieve established objectives; can we use our experience with legal issues to help guide us to sustainability?

19. To what extent are economic systems resulting in ecological risks, including that of our own extinction? How do we deal with those that are posing such problems? Will it be possible for economic forces, views, and paradigms to be seen as human constructs subject to the constraints of reality? When will we realize that sustainable economic growth is an oxymoron? When the human species goes extinct, all monetary economics goes with it to leave the systemic economy intact; reality may be different but it will still be there. The effects of economic influence are among the e_i in Figure 1.4 to result in seeing what we see in human abnormality in Chapter 6. We know the influence exists and have the emotional manifestation of it when we think of the consequences of actions such as reducing the harvest of fish.

20. How pathological, unrealistic, or misleading are our world views as they develop in what is increasingly an environment that is of human design; how much of nature-deficit disorder is a matter of a degraded form of nature? Would it be better if we were personally experiencing more of natural systems as superficially sampled

in therapeutic work in natural settings (Outward Bound, vision quests, ecopsychology), or with animals or horticulture (pet assisted therapy, horticultural therapy)? What are the habits we need so as to result in being a species that is sustainable? How extensive, interrelated, and pathological is control in our thinking, behavior, and habits as manifested in personal, family, business, economic, and ecological realms? How do we deal with our emotional reaction to knowing that saving the lives of children (and the practice of medicine in a more general sense) helps exaggerate the problem of overpopulation? How do we deal with the emotional reaction of knowing that our practice of agriculture, medicine, and most humanitarian aid exaggerates the problem of overpopulation?

Will hope prevail in avoiding our extinction? The overwhelming odds are basis for real pessimism, yet humans are behaviorally extremely adaptable. Can the changes be made? What are the relationships between hope and controlling behaviors? Is psychological change of primary importance, or is it just another critical element that will not work alone?

21. What kind of governments are required to deal with issues so pervasive, extensive, and extreme as overpopulation, use of water, and appropriation of net primary productivity that are measured in orders of magnitude (overpopulation might be measured on a "Malthus scale" as we do earthquakes on a "Richter scale")? How can government be, not just reinvented, but redesigned entirely so as to meet needs that are identified as important for governmental action in bringing about change, whether by government leaders themselves, the people within society, or a combination of both?

22. Different geographic regions and the peoples in them overlap or contain different biotic regions, ecosystems, and physical habitat. Which ones should be designated as human habitat? How do the religious, ethnic, political, and economic issues of such questions get addressed? The capacity to sustain human populations differs from habitat to habitat, country to country (recall Appendix Fig. 6.5.1). How do we resolve conflicts between nations with different resources, different populations, and different levels of sustainability for human populations?

23. Scientists would be advised to present themselves as experts only in the sense of their knowledge of patterns consonant with specific management questions. Questions posed by managers would be answered: "I don't know, but the consonant pattern is...". More specifically, when asked what the appropriate body temperature for humans is, experts would reply "I can't know independent of direct observation, but for a three year old girl, at 9:00 am, in a 40°C environment, immediately after running 100 meters it usually runs about 38°C". When asked what the sustainable harvest of walleye pollock of the eastern Bering Sea might be, experts might indicate: "We can't derive the answer other than through our best estimates of consumption realized for other species. The empirical information indicates that for species with the life history strategy of pollock, nonhuman predators otherwise similar to humans generally consume less than 2% of the standing stock biomass. We need more data for such consumption rates under existing climatological circumstances." Empirical observations must involve measurements of what is addressed in the question. We must ensure that such observations are of the same logical types that are involved in the processes and levels of biological complexity being addressed and specified by the question.

24. This is paralleled at the ecosystem level. Changes by individuals within our species results in species-level changes. Thus, part of what happens is a result of collective effort through the control exhibited by each individual. If we as a species make the changes indicated as necessary by species-level patterns, the rest of the system will respond in ways over which we have no control. But there is a collective response and it requires a collective effort insofar as we have control.

25. Temporal scales enter here in that we cannot make the changes overnight. Individual level (mechanical or short-term) change sufficient to make up for the population-level contribution to the problem of overconsumption (whether it be biomass or energy) would require that everyone cut their consumption to levels that would result in death. This being out of the question as a single element for solving the problem requires change at the population level with a different change for individuals—reduced birth rates or increased mortality. This involves a longer time frame—that associated with species or population level processes.

We are forced to deal with time scale, in part, because it would not be realistic to require individuals to do abnormal (evaluated by comparison with other species—not just in regard to recent history within our species) things to achieve management (even though systemic feedback may involve what we might call abnormal mortality, for example). Thus, to achieve a realistic rate of energy consumption by our species cannot be accomplished by asking each individual human to consume only a stick of celery as their total daily allotment of energy.

26. Such efforts would involve bringing to the attention of the world the fact than excessive CO_2 production is a much larger problem than currently recognized (e.g., reducing CO_2 by 2% per year would take almost 450 years to maximize the information index for CO_2

production—assuming other species won't respond to being relieved of this and other abnormal human influence; Fowler 2008). The other ways we find ourselves to be having pathological influences deserve similar attention and could use the formation of corresponding advocacy groups—all with consistent goals. In the end, everyone has to go beyond advocacy and be part of the change.

27. Keep in mind that abnormal is here defined not simply with reference to current or past human standards, but by comparison to other species as well. Reducing CO_2 production by 80% (as is often expressed as a desirable goal in current efforts to face and deal with the problem) would have to be repeated over five times to bring human production in line with CO_2 production by other species (to maximize diversity, Fowler 2008). Noningested energy consumption by nonhuman species at our trophic level is essentially zero; biomimicry of this example of sustainability will probably seem ludicrous to many. With such examples, however, we begin to appreciate what we have become as a species.

28. The conditions contributing to a raised risk of major pandemic (see "The coming plague" by Garrett 1994), involve much more than exposure to larger numbers of disease organisms as we occupy more and more ecosystems. Counting among the contributing factors is the human population density, our compromised immune systems (e.g., from the effects of toxins), our mobility, genetically inherited immune deficiency in lives saved by medicine, and the reduced resistance brought on by sedentary lives. The genetic effects of our current attempts to "control" disease produces strains or species even more likely to break through what natural immunity, and artificial chemical defenses, we have left. In terms of catastrophe theory, we have developed the conditions for a "perfect storm". Much if this situation is attributable to action carried out in the interest of what are considered good for individuals, societies, or even the short-term interests of our species. Extending consideration to the interests of the nonhuman and long-term interests of our species is part of systemic management.

29. An analogy that may be helpful here is that of the philosophy of Asian martial arts. Central to much of this philosophy is the concept of working *with* forces rather than struggling *against* them (Watts 1951). There are clearly limiting and controlling forces by which ecosystems affect species, including humans. It would seem advisable to move from a view that focuses on a perceived adversarial component to these forces, to a view wherein there is useful and helpful content to information we gather from these forces and their manifestations. As indicated by H. Smith (1994), "...the concept of *wu wei*.... translates literally as inaction but in Taoism means pure effectiveness. Action in the mode of *wu wei* is action in which friction—in interpersonal relationships, in interpsychic conflict, and in relation to nature—is reduced to a minimum."

30. Use of the term "life" is not restricted to the human experience often called life, especially as experienced simply as individuals. It refers to this and comparable aspects of all living systems at all levels of organization, including those that may be intermediate to and extend beyond the spectrum from cells and individuals through species to ecosystems and beyond. It includes all beings with which we share this planet and all biotic systems in which we find ourselves to be a part.

List of Appendices

The following is a list of 22 appendices for the first six chapters of this book (there are none associated with the Epilogue). All are available online (listed according to Appendix number) at: http://www.afsc.noaa.gov/Publications/misc_pdf/Fowler_appendices.pdf.

References

Aarssen, L.W, B.S. Schamp, and J. Pither. 2006. Why are there so many small plants? Implications for species coexistence. *Journal of Ecology* 94: 568–580.

Agee, J.K. and D.R. Johnson (eds). 1988. *Ecosystem management for parks and wilderness.* University of Washington Press, Seattle, WA.

Ahl, V. and T.F.H. Allen. 1996. *Hierarchy theory.* Columbia University Press, New York, NY.

Alexander, R.D. and G. Borgia. 1978. Group selection, altruism and the levels of organization of life. *Annual Review of Ecology and Systematics* 9: 449–474.

Allee, W.C., A.E. Emerson, O. Park, T. Park and K. Schmidt. 1949. *Principles of animal ecology.* W.P. Saunders Co., Philadelphia, PA.

Allen, T.F.H. and T.W. Hoekstra. 1992. *Toward a unified ecology.* Columbia University Press, New York, NY.

Allen, T.F.H. and T.B. Starr. 1982. *Hierarchy: perspectives for ecological complexity.* University of Chicago Press, Chicago, IL.

Allmon, W.D. and R.M. Ross. 1990. Specifying causal factors in evolution: the paleontological contribution. In Ross, R.M. and W.D. Allmon (eds). *Causes of evolution: a paleontological perspective*, pp. 1–17. University of Chicago Press, Chicago, IL.

Alverson, D.L., M.H. Freeberg, J.G. Pope, and S.A. Murawski. 1994. A global assessment of fisheries bycatch and discards. FAO Fisheries Technical Paper. No. 339. RAO, Rome.

American Psychiatric Association. 1980. *Diagnostic and statistical manual of mental disorders.* American Psychiatric Association, Washington, DC.

Anderson, D.J. and J. Kikkawa. 1986. Development of concepts. In Kikkawa, J. and D.J. Anderson (eds). 1986. *Community ecology: pattern and process*, pp. 3–16. Blackwell Scientific Publications, Cambridge, MA.

Anderson, J.E. 1991. A conceptual framework for evaluating and quantifying naturalness. *Conservation Biology* 5: 347–352.

Anderson, S. 1977. Geographic ranges of North American terrestrial mammals. *American Museum Novitates* 2629: 1–15.

Angermeier, P.L. and J.R. Karr. 1994. Biological integrity versus biological diversity as policy directives—protecting biotic resources. *Bioscience* 44: 690–697.

Angliss, R.P. and K.L. Lodge. 2004. Alaska marine mammal stock assessments, 2003. NOAA Techinical Memorandum, NMFS-AFSC-144. U.S. Department of Commerce, Seattle, WA.

Aplet, G.H., N. Johnson, J.T. Olson, and V.A. Sample (eds). 1993. *Defining sustainable forestry.* Island Press, Washington, DC.

Apollonio, S. 1994. The use of ecosystem characteristics in fisheries management. *Reviews in Fisheries Science* 2: 157–180.

Appleman, P. 1970. *Darwin.* Norton, New York, NY.

Arkema, K.K, S.C. Abramson, and B.M. Dewsbury. 2006. Marine ecosystem-based management: from characterization to implementation. *Frontiers in Ecology and the Environment* 4: 525–532.

Arnold, A.J. and K. Fristrup. 1982. The theory of evolution by natural selection: a hierarchical expansion. *Paleobiology* 8: 113–129.

Ashford, R.W. and W. Crewe. 2003. *The parasites of Homo sapiens: an annotated checklist of the protozoa, helminths and arthropods for which we are home.* Taylor and Francis, London.

Backus, R.H. and D.W. Bourne (eds). 1987. *Georges Bank.* MIT Press. Cambridge, MA.

Bakker, R.T. 1983. The deer flees, the wolf pursues: incongruencies in predator–prey coevolution. In Futuyma, D.J. and M. Slatkin (eds). *Coevolution*, pp. 350–382. Sinauer Associates, Sunderland, MA.

Barrett, J.A. 1983. Plant-fungus symbioses. In Futuyma, D.J. and M. Slatkin (eds). *Cooevolution*, pp. 137–160. Sinauer Associates, Sunderland, MA.

Bartholomew, G.A. 1986. The role of natural history in contemporary biology. *Bioscience* 36: 324–329.

Bascompte, J., C.J. Melián, and E. Sala. 2005. Interaction strength combinations and the overfishing of a marine food web. *Proceedings of the National Academy of Sciences of the USA* 102: 5443–5447.

Basset, Y. and R.L. Kitching. 1991. Species number, species abundance, and body length of arboreal arthropods associated with an Australian rainforest tree. *Ecological Entomology* 16: 391–402.

Bateson, G. 1972. Conscious purpose versus nature. In Bateson, G. *Steps to an ecology of mind*, pp. 426–439. Chandler Publishing Co., San Francisco, CA.

Bateson, G. 1979. *Mind and nature: A necessary unity.* Dutton, New York, NY.

Bateson, G. and M.C. Bateson. 1987. *Angels fear: Towards an epistemology of the sacred.* McMillan Publishing Co., New York, NY.

Beattie, M. 1987. *Codependent no more: How to stop controlling others and start caring for yourself.* Harper & Row, New York, NY.

Begon, M., J.L. Harper, and C.R. Townsend. 1990. *Ecology, individuals, populations and communities.* Blackwell Scientific Publications, Cambridge, MA.

Belgrano, A. and C.W. Fowler. 2008. Ecology for management: pattern-based policy. In Munoz, S.I. (ed.). *Ecology Research Progress*, pp. 5–31. Nova Science Publishers, Hauppauge, NY.

Belgrano, A., A.P. Allen, B.J. Enquist, and J.F. Gillooly. 2002. Allometric scaling of maximum population density: a common rule for marine phytoplankton and terrestrial plants. *Ecology Letters* 5: 611–613.

Bell, J.S. 1964. On the Einstein–Podolsky–Rosen paradox. *Physics* 1: 195–200.

Belsky, M.H. 1995. Implementing the ecosystem management approach: optimism or fantasy? *Ecosystem Health* 1: 214–221.

Benton, M.J. and R.J. Twitchett. 2003. How to kill (almost) all life: the end-Permian extinction event. *Trends in Ecology and Evolution* 18: 358–365.

Benyus, J.M. 2002. *Biomimicry: innovation inspired by nature.* Harper Perennial, New York, NY.

Berenbaum, M.R. 1996. Introduction to the symposium: on the evolution of specialization. *American Naturalist* 148: S78–S83.

Berger, J.J. (ed.). 1990. *Environmental restoration: science and strategies for restoring the earth.* Island Press, Washington, DC.

Berry, T. 1997. The universe is our university. *Timeline* 34: 18–23.

Berry, T. 1999. *The great work: our way into the future.* Bell Tower, New York, NY.

Berry, W. 2001. *Life is a miracle.* Counwterpoint, Washington, DC.

Berry, W. and N. Wirzba. 2002. *The art of the commonplace: the agrarian essays of Wendell Berry.* Counterpoint, Washington, D.C.

Blackburn, T.M. and K.J. Gaston. 2006. There's more to macroecology than meets the eye. *Global Ecology and Biogeography* 15: 537–540.

Blackburn, T.M., V.K. Brown, B.M. Doube, J.J.D. Greedwood, J.H. Lawton, and N.E. Stork. 1993. The relationship between abundance and body size in natural animal assemblages. *Journal of Animal Ecology* 62: 519–528.

Blackwelder, R. and G.S. Garoian. 1986. *CRC handbook of animal diversity.* CRC Press, Boca Raton, FL.

Blueweiss, L., H. Fox, V. Kadzma, D. Nakashima, R. Peters, and S. Sams. 1978. Relationships between body size and some life history parameters. *Oecologia* 37: 257–272.

Bock, C.E. 1984a. Geographical correlates of abundance vs. rarity in some North American songbirds: a positive correlation. *American Naturalist* 122: 295–299.

Bock, C.E. 1984b. Geographical correlates of abundance vs. rarity in some North American winter land birds. *Auk* 101: 266–277.

Bock, C.E. 1987. Distribution–abundance relationships of some North American landbirds: a matter of scale? *Ecology* 68: 124–129.

Bock, C.E. and R.E. Ricklefs. 1983. Range size and local abundance of some North American songbirds, a positive correlation. *American Naturalist* 122: 295–299.

Bogan, C.E. and M.J. English. 1994. *Benchmarking for best practices.* McGraw-Hill, Inc., New York, NY.

Bonner, J.T. 1968. Size change in development and evolution. In Macurda, D.B. *Paleobiological aspects of growth and development (Memoir 2)*, pp. 1–15. The Paleontological Society, Menosha, WI.

Botkin, D.B. 1990. *Discordant Harmonies.* Oxford University Press, New York, NY.

Boulter, M. 2002. *Extinction: Evolution and the end of man.* Columbia University Press, New York, NY.

Boxwell, R.J., 1994. *Benchmarking for competitive advantage.* McGraw-Hill, Inc., New York, NY.

Brandon, R.N. 1988. Levels of selection: a hierarchy of interactors. In Plotkin, H.C. (ed.). *The role of behavior in evolution*, pp. 51–71. MIT Press, Cambridge, MA.

Brasier, M.D. 1988. Foraminiferid extinction and ecological collapse during global biological events. In Larwood, G.P. (ed.). *Extinction and survival in the fossil record* pp. 37–64. Oxford University Press, New York, NY.

Bratton, S.P. 1992. Alternative models of ecosystem restoration. In Costanza, R., B.G. Norton, and B.D. Haskell (eds). *Ecosystem health: New goals for environmental management*, pp. 170–189. Island Press, Washington, DC.

Briggs, D.E.G. 1988. Extinction and the fossil record of the arthropods. In Larwood, G.P. (ed.). *Extinction and survival in the fossil record*, pp. 171–209. Oxford University Press, New York, NY.

Brooker, R.W., J.M.J. Travis, E.J. Clark, and C. Dytham. 2007. Modelling species' range shifts in a changing climate: The impacts of biotic interactions, dispersal distance and the rate of climate change. *Journal of Theoretical Biology* 245: 59–65.

Brooks, J.L. 1972. Extinction and the origin of organic diversity. In E.S. Deevey (ed.). *Growth by intussusception; ecological essays in honor or G. Evelyn Hutchinson*, pp. 19–56. Transactions of the Connecticut Academy of Arts and Sciences 44: 1–433.

Brosnan, D.M. and M.J. Groom. 2006. The integration of conservation science and policy: the pursuit of knowledge needs the use of knowledge. In Groom, M.J, G.K. Meffe, and C.R. Carroll (eds). *Principles of conservation biology*, pp. 625–659. Sinauer Associates, Sunderland, MA.

Brown, J.H. 1971. Mammals on mountaintops: nonequilibrium insular biogeography. *American Naturalist* 105: 467–478.

Brown, J.H. 1981. Two decades of homage to Santa Rosalia: toward a general theory of diversity. *American Zoologist* 21: 877–888.

Brown, J.H. 1995. *Macroecology*. University of Chicago Press, Chicago, IL.

Brown, J.H. and B.A. Maurer. 1987. Evolution of species assemblages: effects of energetic constraints and species dynamics on the diversification of the North American avifauna. *American Naturalist* 130: 1–17.

Brown, J.H. and P.F. Nicoletto. 1991. Spatial scaling of species composition: body masses of North American land mammals. *American Naturalist* 138: 1478–1512.

Brown, J.H., P.A. Marquet, and M.L. Taper. 1993. Evolution of body size: consequences of an energetic definition of fitness. *American Naturalist* 142: 573–584.

Brown, J.H., G.C. Stevens, and D.M. Kaufman. 1996. The geographic range: size, shape, boundaries, and internal structure. *Annual Review of Ecology and Systematics* 27: 597–623.

Brown, J.S. and A.O. Parman. 1993. Consequences of size-selective harvesting as an evolutionary game. In Law, R., J.M. McGlade, and T.K. Stokes (eds). *The exploitation of evolving resources: Proceedings of an international conference held at Julich*, pp. 248–261, Germany, Sept. 3–5, 1991. (Lecture notes in biomathematics, 99), Springer-Verlag, Berlin.

Brown, L.R. 1971. Food supplies and the optimum level of population. In Singer, S.F. (ed.). *Is there an optimum level of population?* pp. 72–88. McGraw-Hill Publishing Co., New York, NY.

Brown, R.M. 1986. *The essential Reinhold Niebuhr: Selected essays*. Yale University Press, New Haven, CT.

Brussard, P.F. 1993. Matters of scale. *Science* 262: 1081.

Bulmer, M.G. 1975. A statistical analysis of density dependence. *Biometrics* 31: 901–911.

Burns, T.P., B.C. Patten, and M. Higashi. 1991. Hierarchial evolution in ecological networks: environs and selection. In M. Higashi and T. Burns (eds). *Theoretical studies of ecosystems—the network prospective*, pp. 211–239, Cambridge University Press, New York, NY.

Buss, L.W. 1988. *The evolution of individuality*. Princeton University Press, Princeton, NJ.

Cairns, J. Jr. 1986. The myth of the most sensitive species. *Bioscience* 36: 670–672.

Cairns, J. Jr. 1991. The need for integrated environmental systems management. In Cairns, J. Jr. and T.V. Crawford (eds). *Integrated environmental management*, pp. 5–20. Lewis Publishers, Chelsea, MI.

Calder, W.A., III. 1984. *Size, function and life history*. Harvard University Press, Cambridge, MA.

Calle, E.E., M.J. Thun, J.M. Petrelli, C. Rodriguez, and C.W. Heath. 1999. Body-mass index and mortality in a prospective cohort of U.S. adults. *New England Journal of Medicine* 341: 1097–1105.

Callicott, J.B. 1992. Aldo Leopold's metaphor. In Costanza, R., B.G. Norton, and B.D. Haskell (eds). *Ecosystem health: New goals for environmental management*, pp. 42–56. Island Press, Washington, DC.

Callicott, J.B. 2006. Conservation values and ethics. In Groom, M.J, G.K. Meffe, and C.R. Carroll (eds). *Principles of conservation biology*, pp. 111–135. Sinauer Associates, Sunderland, MA.

Camazine, M., J.-L. Deneubourg, N.R. Franks, J. Sneyd, G. Theraulaz, and E. Bonabeau. 2001. *Self-organization in biological systems*. Princeton University Press, Princeton, NJ.

Camp, R.C. 1995. *Business process benchmarking: finding and implementing best practices*. ASQC Quality Press, Milwaukee, WI.

Campbell, D.T. 1974. 'Downward causation' in hierarchically organized biological systems. In Ayala, F.J. and T. Dobzhansky (eds). *Studies in Philosophy of biology*, pp. 179–185. University Cal. Press, Berkeley, CA.

Campbell, T.M. and T.W. Clark. 1981. Colony characteristics and vertebrate associates of white-tailed and black-tailed prairie dogs in Wyoming. *American Midland Naturalist* 105: 269–276.

Canter, L.W. 1996. *Environmental impact assessment*. McGraw-Hill, Boston, MA.

Capra, F. 1982. *The turning point: science society, and the rising culture*. Simon and Schuster, New York, NY.

Carpenter, S.R. 2002. Ecological futures: building an ecology of the long now. *Ecology* 83: 2069–2083.

Carse, J.P. 1986. *Finite and infinite games*. The Free Press, New York, NY.

Catton, W.R., Jr. 1980. *Overshoot: The ecological basis of revolutionary change*. University of Illinois Press, Chicago, IL.

Caughley, G. 1981. What do we not know about the dynamics of large mammals? In Fowler, C.W. and T.D. Smith (eds). *Dynamics of large mammal populations*, pp. 361–372. Wiley, New York, NY.

Cawte, J. 1978. Gross stress in small islands: a study in macropsychiatry. In Laughlin Jr., C.D. and I.A. Brady (eds). *Extinction and survival in human populations*, pp. 95–121. Columbia Press, Irvington on Hudson, New York.

Ceballos, G. and P. Ehrlich. 2002. Mammal population losses and the extinction crisis. *Science* 296: 904–907.

Chaloner, W.G. and A. Hallam (eds). 1989. *Evolution and extinction: proceedings of a joint symposium of the Royal Society and Linnean Society*, held on 9 and 10 November 1989; The Royal Society of London. Philosophical Transactions of the Royal Society of London 325.

Charlesworth, B., R. Lande, and M. Slatkin. 1982. A neo-Darwinian commentary on macroevolution. *Evolution* 36: 474–498.

Charnov, E.L. 1993. *Life history invariants*. Oxford University Press, New York, NY.

Charnov, E.L., R. Warne, and M. Moses. 2007. Lifetime reproductive effort. *American Naturalist*, E-Article, 170: E129–E142.

Chew, R.M. 1965. Water metabolism of mammals. In Mayer, W.V. and R.G. Van Gelder (eds). *Physiological Mammalogy*, Vol 2. pp. 43–178. Academic Press, New York.

Chiras, D.D. 1992. *Lessons from nature*. Island Press, Washington, DC.

Christensen, N.L., A.M. Bartuska, J.H. Brown, *et al.* 1996. The report of the Ecological Society of America Committee on the scientific basis for ecosystem management. *Ecological Applications* 6: 665–691.

Clark, M.E. 1989. *Ariadne's thread*. St. Martin's Press, New York, NY. 584pp.

Clark, T.W. and D. Zaunbrecher. 1987. The Greater Yellowstone Ecosystem: the ecosystem concept in natural resource policy and management. *Renewable Resources Journal* 5: 8–16.

Clark, T.W., E.D. Amato, D.G. Whittemore, and A.H. Harvey. 1991. Policy and programs for ecosystem management in the Greater Yellowstone Ecosystem: an analysis. *Conservation Biology* 5: 412–422.

Clayton, P. 2004. *Mind and emergence: From quantum to consciousness*. Oxford University Press, New York, NY.

Clinchy, B. M. Belenky, N.R. Goldberger, and J.M. Tarule. 1996. *Knowledge, difference and power: Essays inspired by women's ways of knowing*. Basic Books.

Cohen, J.E. 1995a. Population growth and Earth's human carrying capacity. *Science* 269: 341–346.

Cohen, J.E. 1995b. *How many people can the earth support?* Norton, New York, NY.

Cohen, J.E. 1997. Population, economics, environment and culture: an introduction to human carrying capacity. *Journal of Applied Ecology* 34: 1325–1333.

Cohen, J.E. and C.M. Newman. 1988. Dynamic basis of food web organization. *Ecology* 69: 1655–1664.

Cohen, J.E., F. Briand, and C.M. Newman. 1986. A stochastic theory of community food webs III. Predicted and observed lengths of food chains. *Proceedings of the Royal Society of London, Series B: Biological Sciences* 228: 317–353.

Colborn, T., D. Dumanoski, and J.P. Myers. 1997. *Our stolen future: are we threatening our fertility, intelligence, and survival?: A scientific detective story*. Penguin Group, New York, NY.

Collins, S.L. and S.M. Glenn. 1990. A hierarchical analysis of species' abundance patterns in grassland vegetation. *American Naturalist* 135: 633–648.

Commoner, B. 1971. Survival in the environmental–population crises. In Singer, S.F. (ed.). *Is there an optimum level of population?*, pp. 96–113. McGraw-Hill, New York, NY.

Connell, J.H. and W.P. Sousa. 1983. On the evidence needed to judge ecological stability or persistence. *American Naturalist* 121: 789–824.

Conover, D.O. and S.B. Munch. 2002. Sustaining fisheries yields over evolutionary time scales. *Science* 297: 94–96.

Coombs, I. and D.W.T. Crompton. 1991. *A guide to human helminths*. Taylor and Francis Inc., New York, NY.

Corkett, C.J. 2005. The Pew Report on US fishery councils: a critique from the Open Society. *Marine Policy* 29: 247–253.

Costanza, R. 1992. Toward an operational definition of ecosystem health. In Costanza, R., B.G. Norton, and B.D. Haskell (eds). *Ecosystem health: New goals for environmental management*, pp. 239–256. Island Press, Washington, DC.

Costanza, R. 1995. An unbalanced debate. *Bioscience* 45: 633–634.

Costanza, R., B.G. Norton, and B.D. Haskell (eds). 1992. *Ecosystem health: new goals for environmental management*. Island Press, Washington, DC.

Costanza, R., L.J. Graumlich, and W. Steffen (eds). 2007. *Sustainability or collapse? An integrated history and future of people on earth*. MIT Press, Cambridge, MA.

Cowan, G.A., D. Pines, and D. Melzer (eds). 1994. *Complexity: metaphors, models, and reality*. Santa Fe Institute Studies in the Science of Complexity, Proceedings Volume XVIII. Addison-Wesley, Reading, MA.

Cracraft, J. 1982. A non-equilibrium theory for the rate-control of speciation and extinction and the origin of macroevolutionary patterns. *Systematic Zoology* 31: 348–365.

Cracraft, J. 1985a. Biological diversification and its causes. *Annals of the Missouri Botanical Garden* 72: 794–822.

Cracraft, J. 1985b. Species selection, macroevolutionary analysis and the "hierarchical theory" of evolution. *Systematic Zoology* 34: 222–229.

Cracraft, J. and F.T. Grifo (eds). 1999. *The living planet in crisis: Biodiversity science and policy.* Columbia University Press, New York, NY.

Crawford, R.J.M., P.G. Ryan, and A.J. Williams 1991. Seabird consumption and production in the Benguela and Western Agulhas ecosystems. *South African Journal of Marine Science* 11: 357–375.

Cristoffer, C. 1990. Nonrandom extinction and the evolution and conservation of continental mammal faunas. Ph.D. Dissertation, University of Florida, Gainesville, FL.

Crosby, A.W. 1986. *Ecological imperialism: the biological expansion of Europe, 900–1900.* Cambridge University Press, New York, NY.

Czech, B. 2000. *Shoveling fuel for a runaway train: errant economists, shameful spenders, and a plan to stop them all.* University of California Press, Berkeley, CA.

Czech, B. 2006. If Rome is burning, why are we fiddling? *Conservation Biology* 29: 1563–1565.

Damuth, J.D. 1981. Population density and body size in mammals. *Nature* 290: 699–700.

Damuth, J.D. 1985. Selection among "species": a formulation in terms of natural functional units. *Evolution* 39: 1132–1146.

Damuth, J.D. 1987. Interspecific allometry of population density in mammals and other animals: the independence of body mass and population energy-use. *Biological Journal of the Linnnean Society* 31: 193–246.

Damuth, J.D. 1991. Of size and abundance. *Nature* 351: 268–269.

Damuth, J.D. 1992. Taxon-free characterization of animal communities. In Behrensmeyer, A.K., J.D. Damuth, W.A. DiMichele, R. Potts, H.-D. Sues, and S.L. Wing (eds). *Terrestrial ecosystems through time: evolutionary paleoecology of terrestrial plants and animals,* pp. 184–203. University of Chicago Press, Chicago, IL.

Damuth, J.D. 2007. A macroevolutionary explanation for energy equivilance in the scaling of body size and population density. *American Naturalist* 169: 621–631.

Darlington, P.J. 1971. Nonmathematical models for evolution of altruism, and for group selection. *Proceedings of the National Academy of Sciences of the USA* 69: 293–297.

Darwin, C. 1896. *The origin of species by means of natural selection,* Vols. I and II. D. Appleton and Co., New York, NY.

Darwin, C.G. 1953. *The next million years.* Doubleday & Co., Garden City, NY.

Davies, R.G., C.D.L. Orme, V. Olson, G, *et al.* 2006. Human impacts and the global distribution of extinction risk. *Proceedings of the Royal Society of London, Series B* 273: 2127–2133.

Davis, W.S. and T.P. Simon 1994a. Introduction. In Davis, W.S.and T.P. Simon (eds). *Biological assessment and criteria: tools for water resource planning and decision making* pp. 3–6. Lewis Publishers, Boca Raton, FL.

Davis, W.S. and T.P. Simon (eds). 1994b. *Biological assessment and criteria: tools for water resource planning and decision making.* Lewis Publishers, Boca Raton, FL.

Day, D. 1981. *The doomsday book of animals.* Viking Press, New York, NY.

Dayton, P.K. 1998. Reversal of the burden of proof in fisheries management. *Science* 279: 821–822.

De Vries, H. 1905. *Species and varieties, their origin by mutation.* Open Court, Chicago, IL.

Dell, P.F. and H.A. Goolishian. 1981. Order through fluctuation: An evolutionary epistemology for human systems. *Australian Journal of Family Therapy* 2: 175–184.

Dial, K.P. and J.M. Marzluff. 1988. Are the smallest organisms the most diverse? *Ecology* 69: 1620–1624.

Dial, K.P. and J.M. Marzluff. 1989. Nonrandom diversification within taxonomic assemblages. *Systematic Zoology* 38: 26–37.

Diamond, J.M. 1986. The environmentalist myth. *Nature* 324: 19–20.

Diamond, J.M. 2004. *Collapse: How societies choose to fail or succeed.* Viking Books, New York, NY.

Dickerson, J.E., Jr. and J.V. Robinson. 1986. The controlled assembly of microcosmic communities: the selective extinction hypothesis. *Oecologia* 71: 12–17.

Dobzhansky, T. 1958. Evolution at work. *Science* 127: 1091–1098.

Donaldson, R.J. (ed.). 1979. *Parasites and western man.* University Park Press, Baltimore, MD.

Duarte, C.M., S. Agusti, and H. Peters. 1987. An upper limit to the abundance of aquatic organisms. *Oecologia* 74: 272–277.

Dunstan, J.C., and Jope, K.L. 1993. Pushing the Limits of Boundaries. *The George Wright FORUM* 10: 72–78.

Ecosystem Principles Advisory Panel. 1998. Ecosystem-based fishery management. National Marine Fisheries Service Ecosystem Principles Advisory Panel's Report to Congress, July 6, 1998.

Ehrenfeld, D.J. 1992. Ecosystem health and ecological theories. In Costanza, R., B.G. Norton, and B.D. Haskell (eds). *Ecosystem health: new goals for environmental management,* pp. 135–143. Island Press, Washington, DC.

Ehrenfeld, D.J. 1993. *Beginning again: People and nature in the new millennium.* Oxford University Press, New York, NY.

Ehrenfeld, D.W. 1981. *The arrogance of humanism.* Oxford University Press, New York, NY.

Ehrlich, P.R. 1968. *The population bomb.* Balantine Books, New York, NY.

Ehrlich, P.R. 1980. The strategy of conservation, 1980–2000. In Soulé M.E. and M.A. Wilcox (eds). *Conservation biology, an evolutionary-ecological perspective,* pp. 329–344. Sinauer Associates, Sunderland, MA.

Ehrlich, P.R. 1985. Ecosystems and ecosystem functions: implications for humankind. In Hoage, R.J. (ed.). *Animal extinctions; what everyone should know,* pp. 159–173. Smithsonian Institution Press, Washington, DC.

Ehrlich, P.R. 1986. Extinction: what is happening now and what needs to be done. In Elliott D.K. (ed.). *Dynamics of extinction,* pp. 157–164, John Wiley & Sons, New York.

Ehrlich, P.R. 1989. Discussion: Ecology and resource management—is ecological theory any good in practice? In Roughgarden, J., R.M. May, and S.A. Levin (eds). *Perspectives in ecological theory,* pp. 306–318. Princeton University Press, Princeton, NJ.

Ehrlich, P.R. 1991. Population diversity and the future of ecosystems. *Science* 254: 175.

Ehrlich, P.R. 1995. The scale of the human enterprise and biodiversity loss. In Lawton, J.H. and R.M. May (eds). *Extinction rates,* pp. 214–226. Oxford University Press, New York, NY.

Ehrlich, P.R. and A.H. Ehrlich. 1981. *Extinction: the causes and consequences of the disappearance of species.* Random House, New York, NY.

Ehrlich, P.R. and A.H. Ehrlich. 1990. *The population explosion.* Simon and Schuster, New York, NY.

Ehrlich, P.R. and A.H. Ehrlich. 1992. The most over-populated nation. In Grant, L. (ed.). *Elephants in the Volkswagon,* pp. 125–135. W.H. Freeman and Co., San Francisco, CA.

Ehrlich, P.R. and A.H. Ehrlich. 1996. *Betrayal of science and reason.* Island Press, Washington, DC.

Ehrlich, P.R. and J.P. Holdren. 1974. Impact of population growth. *Science* 171: 1212–1217.

Ehrlich, P.R. and H.A. Mooney. 1983. Extinction, substitution, and ecosystem services. *Bioscience* 33: 248–254.

Ehrlich, P.R. and E.O. Wilson. 1991. Biodiversity studies: science and policy. *Science* 253: 758–762.

Eisenberg, J.F. 1981. *The mammalian radiations.* University of Chicago Press, Chicago, IL.

El-Hani, C.N. and C. Emmeche. 2000. On some theoretical grounds for an organism-centered biology: Property emergence, supervenience, and downward causation. *Theory in Biosciences* 119: 234–275.

Eldredge, N. 1985. *Unfinished synthesis: Biological hierarchies and modern evolutionary thought.* Oxford University Press, New York, NY.

Eldredge, N. 1989. *Macroevolutionary dynamics: Species, niches, and adaptive peaks.* McGraw-Hill Publishing Co., New York, NY.

Eldredge, N. 1991. *The miner's canary: Unraveling the mysteries of extinction.* Prentice Hall Press, New York, NY.

Eldredge, N. and J. Cracraft. 1980. *Phylogenetic patterns and the evolutionary process.* Columbia University Press, New York, NY.

Eldredge, N. and S.J. Gould. 1972. Punctuated equilibria: an alternative to phyletic gradualism. In Schopf, T.J.M. (ed.). *Models in paleobiology,* pp. 82–115. Freeman, San Francisco, CA.

Elliott, D.K. (ed.). 1986. *Dynamics of extinction.* John Wiley & Sons, New York, NY.

Emerson, A.E. 1960. The evolution of adaptation in population systems. In Tax, S. (ed.). *The evolution of life,* pp. 307–348. University of Chicago Press, Chicago, IL.

Emlen, J.M., C.C. Freeman, A. Mills, and J.H. Graham. 1998. How organisms do the right thing: the attractor hypothesis. *Chaos* 8: 717–726.

Emmons, C.W., C.H. Binford, U.P. Utz, and K.J. Kwon-Chung. 1977. *Medical mycology* (3rd edn.). Lea & Hebiger, Philadelphia, PA.

Encyclopedia Britannica. 1977. *Helen Hemingway* Benton, Chicago, IL.

Enger, E. and B.F. Smith. 2000. *Environmental science: a study of relationships.* McGraw-Hill, New York, NY.

Enquist, B.J. and K.J. Niklas. 2002. Global allocation rules for patterns of biomass partitionaing in seed plants. *Science* 295: 1517–1520.

Enright, J.T. 1969. Zooplankton grazing rates estimated under field conditions. *Ecology* 50: 1070–1075.

Erwin, T.L. 1982. Tropical forests: their richness in Coleoptera and other arthropod species. *Coleopterists' Bulletin* 36: 74–75.

Erwin, T.L. 1983. Beetles and other insects of tropical forest canopies a Manaus, Brazil, sampled by insecticidal fogging. In Sutton, S.L., T.C. Whitmore, and A.C. Chadwick (eds). *Tropical rain forest: Ecology and management,* pp. 59–75. Blackwell, Oxford.

Etnier, M.A. and C.W. Fowler. 2005. Comparison of size selectivity between marine mammals and commercial fisheries with recommendations for restructuring management policies. NOAA Technical Memorandum NMFS-AFSC-159. U.S. Department of Commerce, Seattle, WA.

Horgan, J. 1999. *The undiscovered mind: How the human brain defies replication, medication, and explanation.* Free Press, New York, NY.

Horibuchi, S. and Y. Inoue (eds). 1994. *Stereogram.* Cadence Books, San Francisco, CA.

Horsefall, F.K., Jr. and I. Tamm (eds). 1965. *Viral and rickettsial infections of man.* J.B. Lippincott Co., Philadelphia, PA.

Howe, H.F. 1977. Bird activity and seed dispersal of a tropical wet forest tree. *Ecology* 58: 539–550.

Howson, C. and P. Urbach. 1991. Bayesian reasoning in science. *Nature* 350: 371–374.

Hubbell, S.P. 2001. *The unified neutral theory of biodiversity and biogeography.* Princeton University Press, Princeton, NJ.

Hugueny, B. 1990. Geographic range of west African freshwater fishes: role of biological characteristics and stochastic processes. *Acta Oecologica* 11: 351–375.

Hull, B. 2006. *Infinite nature.* University of Chicago Press, Chicago, IL.

Hull, D.L. 1976. Are species really individuals? *Systematic Zoology* 25: 174–191.

Hull, D.L. 1978. A matter of individuality. *Philosophy of Science* 45: 335–360.

Hull, D.L. 1980. Individuality and selection. *Annual Review of Ecology and Systematics* 11: 311–332.

Huntley, B.J., E. Ezcurra, E.R. Fuentes *et al.* 1991. A sustainable biosphere: the global perspective. *Ecology International* 20: 1–14.

Hutchinson, G.E. and R.H. MacArthur. 1959. A theoretical ecological model of size distributions among species of animals. *American Naturalist* 93: 117–125.

Inchausti, P. and J. Halley. 2001. Investigating long-term ecological variability using the global population dynamics database. *Science* 293: 655–657.

Interagency Ecosystem Managment Task Force. 1995. *The ecosystem approach: Healthy ecosystems and sustainable economics.* Vol. I—Overview. Report PB95–265575. National Technical Information Services, Springfield VA.

Jablonski, D. 2007. Evolution and the levels of selection. *Science* 316: 1428–1430.

Jacobson, S.K. 1990. Graduate education in conservation biology. *Conservation Biology* 4: 431–440.

Janzen, D.H. 1983. Dispersal of seeds by vertebrate guts. In Futuyma, D.J. and M. Slatkin (eds). *Coevolution,* pp. 232–262. Sinauer Associates, Sunderland, MA.

Janzen, D.H. and P.S. Martin. 1982. Neotropical anachronisms: the fruits the gomphotheres ate. *Science* 215: 19–27.

Jarvis, S. 1978. Extinction by default? A philosophical argument of education to enable survival. Ms. Thesis. University of Washington, Seattle, WA.

Jenkins, R.E. 1985. The identification, acquisition, and preservation of land as a species conservation strategy. In Hoage, R.J. (ed.). *Animal extinctions: What everyone should know,* pp. 129–145. Smithsonian Institution Press, Washington, DC.

Johnson, R.M. 1987. *A logic book.* Wadsworth Publishing Co., Belmont, CA.

Johnston, C.M. 1991. *Necessary wisdom: Meeting the challenge of a new cultural maturity.* ICD Press, Seattle, WA.

Johnston, C.M. 1994. *Pattern & reality: A brief introduction to creative systemss theory.* ICD Press, Seattle, WA.

Jordan, W.R., III, M.E. Gilpin, and J.D. Aber (eds). 1990. *Restoration ecology: A synthetic approach to ecological research.* Cambridge University Press, New York, NY.

Jordano, P. 1987. Patterns of mutualistic interactions in pollination and seed dispersal: connectance, dependence, asymmetries, and coevolution. *American Naturalist* 129: 657–677.

Jørgensen, S.E. 1992. *Integration of ecosystem theories: A pattern.* Kluwer Academic Publishers, Boston, MA.

Jørgensen, S.E. and W.J. Mitsch. 2000. Ecological engineering. In Jørgensen, S.E. and F. Müller, (eds). *Handbook of ecosystem theories and management,* pp. 537–546. Lewis Publishers, Inc., Boda Raton, FL.

Jørgensen, S.E., B.C. Patten, and M. Straskraba. 1999. Ecosystems emerging: 3. Openness. *Ecological Modelling* 117: 41–64.

Juanes, F. 1986. Population density and body size in birds. *American Naturalist* 128: 921–929.

Kangas, P. 2004. *Ecological engineering: Principles and practices.* CRC Press, Boca Raton, FL.

Kareiva, P. 1989. Renewing the dialogue between theory and experiments in population ecology. In Roughgarden, J., R.M. May and S.A. Levin (eds). *Perspectives in ecological theory,* pp. 68–88. Princeton University Press, Princeton, NJ.

Karplus, W.J. 1977. The place of systems ecology models in the spectrum of mathematical models. In Innis, G.S. (ed.). *New directions in the analysis of ecological systems (Part 2),* pp. 225–228. The Society for Computer Simulation, La Jolla, CA.

Karr, J.R. 1982a. Avian extinction on Barro Colorado Island, Panama: a reassessment. *American Naturalist* 119: 220–239.

Karr, J.R. 1987. Biological monitoring and enviornmental assessment: a conceptual framework. *Environmental Management* 11: 249–256.

Karr, J.R. 1990. Avian survival rates and the extinction process on Barro-Colorado Island, Panama. *Conservation Biology* 4: 391–397.

Karr, J.R. 1991. Biological integrity: a long neglected aspect of water resource management. *Ecological Applications* 1: 66–84.

Karr, J.R. 1992. Ecological integrity: protecting earth's life. In Costanza, R., B.G. Norton and B.D. Haskell (eds). *Ecosystem health: New goals for environmental management*, pp. 223–238. Island Press, Washington, DC.

Karr, J.R. 1994. Protecting aquatic ecosystems: clean water is not enough. In Davis, W.S. and T.P. Simon (eds). *Biological assessment and criteria: Tools for water resource planning and decision making*, pp. 7–13. Lewis Publishers, Boca Raton, FL.

Katz, M., D.D. Despommier, and R.W. Gwadz. 1989. *Parasitic diseases*. Springer-Verlag, New York, NY.

Kauffman, S.A. 1993. *The origins of order: Self-organization and selection in evolution*. Oxford University Press, New York, NY.

Keddy, P.A., H.T. Lee, and I.C. Wisheu. 1993. Choosing indicators of ecosystem integrity: wetlands as a model system. In Woodley, S., G. Francis, and J. Kay (eds). *Ecological integrity and the management of ecosystems*, pp. 61–79. St. Luicie Press, Delray Beach, FL.

Keiter, R.B. 1988. Natural ecosystem management in park and wilderness areas: looking at the law. In Agee, J.K. and D.R. Johnson (eds). *Ecosystem management for parks and wilderness*, pp. 15–40. University of Washington Press, Seattle, WA.

Kelt, D.A., and D.H. Van Vuren. 2001. The ecology and macroecology of mammalian home range area. *American Naturalist* 157: 637–645.

Kenchington, R.A. 1990. *Managing marine environments* Taylor and Francis, New York, NY.

Kerfoot, W.C. and A. Sih (eds). 1989. *Predation: Direct and indirect impacts on aquatic communities*. University Press of New England, Hanover, NH.

Kiehl, J.T. and C.A. Shields. 2005. Climate simulation of the latest Permian: implications for mass extinction. *Geology* 33: 757–760.

Kingsland, S.E. 1985. *Modeling nature: Episodes in the history of population ecology*. University of Chicago Press, Chicago, IL.

Kitchell, J.A. 1985. Evolutionary paleoecology: recent contributions to evolutionary theory. *Paleobiology* 11: 91–104.

Kleiber, M. 1961. *The fire of life; an introduction to animal energetics*. John Wiley and Sons, New York, NY.

Knight, R.L. and S.F. Bates (eds). 1995. *A new century for natural resource management*. Island Press, Washington, DC.

Knoll, A.H. 1989. Evolution and extinction in the marine realm: some constraints imposed by phytoplankton. In Chaloner, W.G. and A. Hallam. *Evolution and extinction: Proceedings of a joint symposium of the Royal Society of the Linnean Society* held on 9 and 10 November 1989, pp. 279–290. The Royal Society, London.

Koehl, M.A.R. 1989. Discussion: from individuals to populations. In Roughgarden, J., R.M. May and S.A. Levin (eds). *Perspectives in ecological theory*, pp. 39–53. Princeton University Press, Princeton, NJ.

Koestler, A. 1978. *Janus: A summing up*. Random House, New York, NY.

Koestler, A. 1982. *The ghost in the machine*. McMillan, New York, NY.

Korzybski, A. 1933. *Science and sanity: An introduction to non-Aristotelian systems and general semantics*. International Non-Aristotelian Library, Lancaster, PA.

Kuhn, T.S. 1962. *The structure of scientific revolutions*. University of Chicago Press, Chicago, IL.

Kunkel, K. and D.H. Pletscher. 1999. Species-specific population dynamics of cervids in a multipredator ecosystem. *Journal of Wildlife Management* 63: 1082–1093.

LaBarbera, M. 1989. Analyzing body size as a factor in ecology and evolution. *Annual Review of Ecology and Systematics* 20: 97–117.

Lackey, R.T. 1998. Seven pillars of ecosystem management. *Landscape and Urban Planning* 40: 21–30.

Larkin, P.A. 1977. An epitaph for the concept of maximum sustained yield. *Transactions of the American Fisheries Society* 106: 1–11.

Laughlin, C.D., Jr. and I.A. Brady (eds). 1978. *Extinction and survival in human populations*. Columbia Press, Irvington on Hudson, New York.

Lavigne, D. 2006 (ed.). *Gaining ground: In pursuit of ecological sustainability*. International Fund for Animal Welfare, London, ON, Canada.

Lavigne, D. R.K. Cox, V. Menon, and M. Wamithi. 2006. Reinventing wildlife conservation for the 21st century. In Lavigne, D. (ed.) *Gaining ground: In pursuit of ecological sustainability*, pp. 379–406. International Fund for Animal Welfare, London, ON, Canada.

Law, R. 2001. Phenotypic and genetic changes due to selective exploitation. In Reynolds, J.D., G.M. Mace, K.H. Redford, and J.G. Robinson (eds). *Conservation of exploited species*, pp. 324–342. Cambridge University Press, Cambridge.

Law, R., J.M. McGlade, and T.K. Stokes (eds). 1993. *The exploitation of evolving resources: proceedings of an international conference held at Julich*, Germany, Sept. 3–5, 1991 (Lecture notes in biomathematics, 99). Springer-Verlag, Berlin.

Lawton, J.H. 1974. Review of J.M. Smith's Models in Ecology. *Nature* 248: 537.

Lawton, J.H. 1989a. What is the relationship between population density and body size in animals? *Oikos* 55: 429–434.

and M. Slatkin (eds). *Coevolution*, Sinauer Associates, Sunderland, MA.

Orme, C.D.L., R.G. Davies, V.A. Olson, *et al.* 2006. Global patterns of geographic range size in birds. *PLOS Biology* 4: 1276–1283.

Ormerod, P. 2006. *Why most things fail: Evolution, extinction and economics.* Pantheon, New York, NY.

Overholtz, W.J., S.A. Murawski, and K.L. Foster. 1991. Impact of predatory fish, marine mammals, and seabirds on the pelagic fish ecosystem of the northeastern USA. ICES Marine *Science Symposia* 193: 198–208.

Ovington, J.D. 1975. Strategies for the management of natural and man-made ecosystems. In van Dobben, W.H. and R.H Lowe-McConnell (eds). *Unifying concepts in ecology*, pp. 239–246. Dr. W. Junk b. v., The Hague, the Netherlands, and the Centre for Agricultural Publishing and Documentation, Wageningen, the Netherlands.

Owen-Smith, R.N. 1988. *Megaherbivores: The influence of very large body size on ecology.* Cambridge University Press, New York, NY.

Paddock, W.C. 1971. Agriculture as a force in determining the United States' optimum population size. In Singer, S.F. (ed.). *Is there an optimum level of population?* pp. 89–95 McGraw-Hill Publishing Co., New York, NY.

Page, T. 1992. Environmental existentialism. In Costanza, R., B.G. Norton and B.D. Haskell (eds). *Ecosystem health: New goals for environmental management*, pp. 97–123. Island Press, Washington, DC.

Pagel, M.D., P.H. Harvey, and H.C.J. Godfray. 1991. Species-abundance, biomass, and resource-use distributions. *American Naturalist* 138: 836–850.

Pagel, M.D., R.M. May, and A.R. Collie. 1991. Ecological aspects of the geographical distribution and diversity of mammalian species. *American Naturalist* 137: 791–815.

Paine, R.T. 1988. Food webs: road maps of interactions or grist for theoretical development? *Ecology* 69: 1648–1654.

Paine, R.T. 1992. Food web analysis through field measurement of per capita interaction strength. *Nature* 355: 73–75.

Parenteau, P.A. 1998. Rearranging the deck chairs: endangered species act reforms in an era of mass extinction. *William & Mary Environmental Law and Policy Review.* 22: 227–311.

Park, T. 1948. Experimental studies of interspecies competition. I. Competition between populations of the flour beetles *Tribolium confusum* Duval and *Tribolium castaneum* Herbst. *Ecological Monographs* 18: 265–308.

Parker, A.C. 1916. The consititution of the five nations. New York State Museum Bulletin, No. 184. The University of the State of New York, Albany.

Parsons, P.A. 1991a. Evolutionary rates: stress and species boundaries. *Annual Review of Ecology and Systematics* 22: 1–18.

Parsons, P.A. 1991b. Stress and evolution. *Nature* 30: 356–357.

Parvinen, K. 2005. Evolutionary suicide. *Acta Biotheoretica* 53: 241–264.

Patten, B.C. 1991. Network ecology: indirect determination of the life–environment relationship in ecosystems. In Higashi, M. and T. Burns (eds). *Theoretical studies of ecosystems—the network prospective*, pp. 288–351. Cambridge University Press, New York, NY.

Patten, D.T. 1991. Human impacts in the greater Yellowstone: evaluating sustainability goals and eco-redevelopment. *Conservation Biology* 5: 405–411.

Patterson, B.D. 1984. Mammalian extinction and biogeography in the southern rocky mountains. In Nitecki, M.H. (ed.). *Extinctions*, pp. 247–294. The University of Chicago Press, Chicago, IL.

Pauly, D., V. Christensen, J. Dalsgaard, R. Rroese, and F. Torres Jr. 1998. Fishing down marine food webs. *Science* 279: 860–863.

Pennycuick, C.J. 1992. *Newton rules biology: A physical approach to biological problems.* Oxford University Press, New York, NY.

Perez, M.A. and W.B. McAlister. 1993. *Estimates of food consumption by marine mammals in the Eastern Bering Sea* NOAA Technical Memorandum NMFS-AFSC-14. US Deptartment of Commerce, Seattle, WA.

Petchey, O.L., A.L. Downing, G.G. Mittelbach, L. Persson, C.F. Steiner, P.H. Warren, and G. Woodward. 2004. Species loss and the structure and functioning of multitrophic aquatic systems. *Oikos* 104: 467–478.

Peterman, R.M. 1990. Statistical power analysis can improve fisheries research and management. *Canadian Journal of Fisheries and Aquatic Sciences* 47: 2–15.

Peterman, R.M. and M. M'Gonigle. 1992. Statistical power analysis and the precautionary principle. *Marine Pollution Bulletin* 24: 231–234.

Peters, R.H. 1983. *The ecological implications of body size.* Cambridge University Press, New York, NY.

Peters, R.H. 1991. *A critique for ecology.* Cambridge University Press, New York, NY.

Peters, R.H. and K. Wassenberg. 1983. The effect of body size on animal abundance. *Oecologia* 60: 89–96.

Petraits, P., R.E. Latham, and R.A. Neisenbaum. 1989. The maintenance of species diversity by disturbance. *Quarterly Review of Biology* 64: 393–418.

Phillips, C.G. and J. Randolph. 2000. The relationship of ecosystem management to NEPA and its goals. *Environmental Management* 26: 1–12.

Pielou, E.C. 1969. *An introduction to mathematical ecology.* John Wiley and Sons, New York.

Pilkey, O.H. and L. Pilkey-Jarvis. 2007. *Useless arithmetic: Why environmental scientists can't predict the future.* Colombia University Press, New York, NY.

Pimentel, D. 1966. Complexity of ecological systems and problems in their study and management. In Watt, K.E.F. (ed.). *Systems analysis and ecology,* pp. 15–35. Academic Press, New York, NY.

Pimentel, D. 1986. Agroecology and economics. In Kogen, M. (ed.). *Ecological theory and integrated pest management practice,* pp. 299–320. John Wiley & Sons, New York, NY.

Pimentel, D. and M. Pimentel. 1992. Land energy and water: the constraints governing ideal U.S. population size. In Grant, L. (ed.). *Elephants in the Volkswagon,* pp. 18–31. W.H. Freeman and Co., San Francisco, CA.

Pimentel, D., S.A. Levin, and A.B. Soans. 1975. On the evolution of energy balance in some exploiter-victim systems. *Ecology* 56: 381–390.

Pimentel, D., U. Stachow, D.A. Takacs, H.W. Brubaker, A.R. Dumas, J.J. Meaney, J.A.S. O'Neil, D.E. Onsi, and D.B. Corzilius. 1992. Conserving biological diversity in agricultural/forestry systems: most biological diversity exists in human-managed ecosystems. *Bioscience* 42: 354–362.

Pimentel, D., H. Acquay, M. Biltonen, *et al.* 1992. Environmental and economic costs of pesticide use. *Bioscience* 42: 750–758.

Pimentel, D., H. Acquay, M. Biltonen, *et al.* Assessment of environmental and economic costs of pesticide use. In Pimentel, D. and H. Lehman (eds) *The pesticide question: Environment, economics, and ethics,* pp. 47–84. Chapman and Hall, New York, NY.

Pimentel, D., C. Harvey, P. Resosudarmo, *et al.* 1995. Environmental and economic costs of soil erosion and conservation benefits. *Science* 267: 1117–1123.

Pimentel, D., S. Cooperstein, H. Randell, *et al.* 2007. Ecology of increasing diseases: population growth and environmental degradation. *Human Ecology* 35: 653–668.

Pimm, S.L. 1980. Properties of food webs. *Ecology* 61: 219–225.

Pimm, S.L. 1982. *Food webs.* Chapman and Hall, London.

Pimm, S.L. 1991. *The balance of nature? Ecological issues in the conservation of species in communities.* The University of Chicago Press, Chicago, IL.

Pimm, S.L. and M.E. Gilpin. 1989. Theoretical issues in conservation biology. In Roughgarden, J., R.M. May and S.A. Levin (eds) *Perspectives in ecological theory,* pp. 287–305. Princeton University Press, Princeton, NJ.

Pimm, S.L. and R.L. Kitching. 1988. Food web patterns: trivial flaws or the basis of an active research program? *Ecology* 69: 1669–1672.

Pimm, S.L. and A.J. Redfearn. 1988. The variability of population densities. *Nature* 334: 613–108.

Plotkin, B. 2008. *Nature and the human soul: Cultivating wholeness and community in a fragmented world.* The World Library, Novato, CA.

Policansky, D. 1987. Evolution, sex, and sex allocation. *Bioscience* 37: 466–468.

Ponting, C. 1991. *A green history of the world: The environment and the collapse of great civilizations.* Sinclair-Stevenson, London.

Potter, V.R. 1990. Getting to the year 3000: can global bioethics overcome evolution's fatal flaw? *Perspectives in Biology and Medicine* 34: 89–98.

Power, M.J., D. Tilman, J.A. Estes, *et al.* 1996. Challenges in the quest for keystones. *Bioscience* 46: 609–620.

Pratarelli, M.E. and B. Chiarelli. 2007. Extinction and overspeciallization: the dark side of human innovation. *Mankind Quarterly* 48: 83–98.

Preston, F.W. 1949. The commonness, and rarity, of species. *Ecology* 29: 254–283.

Preston, F.W. 1962. The canonical distribution of commonness and rarity. *Ecology* 43: 185–215.

Prigogine, I. 1978. Time, structure, and fluctuations. *Science* 201: 777–785.

Prigogine, I. and I. Stengers. 1984. *Order out of chaos: Man's new dialogue with nature.* Bantam Books, New York, NY.

Punt, A.E. and A.D.M. Smith. 2001. The gospel of maximum sustainable yield in fisheries management: birth, crucifixion and reincarnation. In Reynolds, J.D., G.M. Mace, K.H. Redford, and J.G. Robinson (eds) *Conservation of exploited species,* pp. 41–66. Cambridge University Press, Cambridge.

Pyle, R.M. 1980. Management of nature reserves. In Soulé, M.E. and M.A. Wilcox (eds) *Conservation biology, an evolutionary-ecological perspective,* pp. 319–328. Sinauer Associates, Sunderland, MA.

Rankin, D.J. and A. Lopez-Sepulcre. 2005. Can adaptation lead to extinction? *Oikos* 111: 616–619.

Rapport, D.J. 1989a. State of ecosystem medicine. *Perspectives in Biology and Medicine* 33: 120–132.

Rapport, D.J. 1989b. What constitutes ecosystem health? *Perspectives in Biology and Medicine* 33: 120–132.

Rapport, D.J. 1989c. Symptoms of pathology in the Gulf of Bothnia (Baltic Sea): Ecosystem response to stress from human activity. *Biological Journal of the Linnnean Society* 37: 33–49.

Rapport, D.J. 1992. What is clinical ecology? In Costanza, R., B.G. Norton and B.D. Haskell (eds). *Ecosystem health: new goals for environmental management*, pp. 144–156. Island Press, Washington, DC.

Rapport, D.J. and R. Moll. 2000. Applications of ecosystem theory and modelling to assess ecosystem health. In Jørgensen, S.E., and Müller, F. (eds), *Handbook of ecosystem theories and management*, pp. 487–496. Lewis Publishers, Boca Raton, FL.

Rapport, D.J., H.A. Regier, and T.C. Hutchinson. 1985. Ecosystem behavior under stress. *American Naturalist* 125: 617–640.

Rapport, D.J., H.A. Regier, and C. Thorpe. 1981. Diagnosis, prognosis, and treatment of ecosystems under stress. In Barrett, G.W. and R. Rosenberg (eds). *Stress effects on natural ecosystems*, pp. 269–280. John Wiley & Sons, New York, NY.

Rasmussen, L.L. 1996. Earth community Earth ethics. Orbis Books, Maryknoll, NY.

Raup, D.M. 1984. Death of species. In Nitecki, M.H. (ed.). *Extinctions*, pp. 1–19 The University of Chicago Press, Chicago, IL.

Raup, D.M. 1986. Biological extinction in earth history. *Science* 231: 1528–1533.

Raup, D.M. and G.E. Boyajian. 1988. Patterns of generic extinction. *Paleobiology* 14: 109–125.

Raven, P.H. and J. Cracraft. 1999. Seeing the world as it really is: global stability and environmental change. In Cracraft, J. and F.T. Grifo (eds). *The living planet in crisis: Biodiversity science and policy*, pp. 287–298. Columbia University Press, New York, NY.

Redman, C.L. 1999. *Human impact on ancient enviornments*. University of Arizona Press, Tucson, AZ.

Reed, C.G. 1989. More on extinction. *Bioscience* 39: 357–358.

Reed, E.S. 1981. The lawfulness of natural selection. *American Naturalist* 118: 61–71.

Rees, W.E. 2002. Globalization and sustainability: conflict or convergence? *Bulletin of Science, Technology and Society* 22: 249–268.

Rees, W.E. and M. Wackernagel. 1996. *Our ecological footprint: Reducing human impact on the Earth*. New Society Publishers, Philadelphia, PA.

Regal, P.J. 1996. Metaphysics in genetic engineering: cryptic philosophy and ideology in the "science" of risk assessment. In A. Van Dommelen (ed.). *Coping with deliberate release: The limits of risk assessment*, pp. 15–32. International Centre for Human and Public Affairs, Tilburg (Netherlands) & Buenos Aires.

Regier, H.A. 1973. Sequence of exploitation of stocks in multispecies fisheries in the Laurentian Great Lakes. *Journal of the Fisheries Research Board of Canada* 30: 1992–1999.

Regier, H.A. 1993. The notion of natural and cultural integrity. in Woodley, S., G. Francis, and J. Kay (eds). *Ecological integrity and the management of ecosystems*, pp. 3–18. St. Luicie press, Delray Beach, FL.

Regier, H.A. and W.L. Hartman. 1973. Lake Erie's fish community: 150 years of cultural stresses. *Science* 180: 1248–1255.

Rensch, B. 1959. *Evolution above the species level*. Columbia University Press, New York, NY.

Restrepo, V.R., G.G. Thompson, P.M. Mace, W.L. Gabriel, L.L. Low, A.D. MacCall, R.D. Methot, J.E. Powers, B.L. Taylor, P.R. Wade, and J.F. Witzig. 1998. *Technical guidance on the use of precautionary approaches to implementing National Standard 1 of the Magnuson-Stevens Fishery Conservation and Management Act*. NOAA Technical Memorandum NMFS-F/SPO-31. US Department of Commerce, Washington, DC.

Ricklefs, R.E. 1989. Speciation and diversity: the integration of local and regional processes. In Otte, D. and J.A. Endler (eds). *Speciation and its consequences*, pp. 599–622. Sinauer Associates, Sunderland, MA.

Ridgway, S.H., and R. Harrison (eds). 1981–1999. *Handbook of marine mammals*, Vol 1–6. Academic Press, New York, NY.

Robinson, J.G. and K.H. Redford. 1986. Body size, diet, and population density of neotropical forest mammals. *American Naturalist* 128: 665–680.

Rockford, L.L., R.E. Stewart, and T. Dietz (eds). 2008. *Foundations of environmental sustainability*. Oxford University Press, New York, NY.

Roe, S.A. 1983. The generative process. *Science* 220: 494.

Roosevelt, D. 2002. *Grandmère: A personal history of Eleanor Roosevelt*. Warner Books, Inc., New York, NY.

Rosen, R. 1987. Some epistemological issues in physics and biology. In Hiley, B.J., and F.D. Peat. *Quantum implications*, pp. 315–327. Routledge & Kegan Paul, New York.

Rosenberg, A. 1985. *The structure of biological science*. Cambridge University Press, New York, NY.

Rosenberg, A.A., M.J. Fogarty, M.P. Sissenwine, J.R. Beddington, J.G. Shephard. 1993. Achieving sustainable use of renewable resources. *Science* 262: 828–829.

Rosenzweig, M.L. 1971. Paradox of enrichment: destabilization of exploitation ecosystems in ecological time. *Science* 171: 385–387.

Rosenzweig, M.L. 1974. *And replenish the earth: The evolution, consequences, and prevention of overpopulation*. Harper & Row, New York, NY.

Rosenzweig, M.L. 1975. On continental steady states of species diversity. In M.L. Cody and J.M. Diamond (eds). *Ecology and evolution of communities*, pp. 121–140. Belknap, Cambridge, MA.

Rosenzweig, M.L. 1995. *Species diversity in space and time.* Cambridge University Press, New York, NY.

Roszak, T. 1992. *The voice of the earth.* Simon and Schuster, New York, NY.

Roszak, T., M.E. Gomes, and A.D. Kanner (eds.). 1995. *Ecopsychology: Restoring the earth, healing the mind.* Sierra Club Books, San Francisco, CA.

Roughgarden, J. 1983. The theory of coevolution. In Futuyma, D.J. and M. Slatkin (eds). *Coevolution*, pp. 33–64. Sinauer Associates, Sunderland, MA.

Roughgarden, J. 1989. The structure and assembly of communities. In Roughgarden, J., R.M. May and S.A. Levin (eds). *Perspectives in ecological theory*, pp. 203–226. Princeton University Press, Princeton, NJ.

Roughgarden, J. 1991. The evolution of sex. *American Naturalist* 138: 934–953.

Ruggiero, A. and B.A. Hawkins. 2006. Mapping macroecology. *Global Ecology and Biogeography* 15: 433–437.

Safina, C. 1995. The world's imperiled fish. *Scientific American* 273: 46–53.

Sagoff, M. 1992. Has nature a good of its own? In Costanza, R., B.G. Norton and B.D. Haskell (eds). *Ecosystem health: New goals for environmental management*, pp. 57–71. Island Press, Washington, DC.

Salt, F.W. 1979. A comment on the use of the term emergent properties. *American Naturalist* 113: 145–148.

Salthe, S.N. 1985. *Evolving hierarchical systems: Their structure and representation.* Columbia University Press, New York, NY.

Salwasser, H., D.W. MacCleery, and T.A. Snellgrove. 1993. An ecosystem perspective on sustainable forestry and new directions for the U.S. national forest system. In Aplet, G.H., N. Johnson, J.T. Olson, and V.A. Sample (eds). *Defining sustainable forestry*, pp. 44–89. Island Press, Washington, DC.

Salzman, L. 1994. Economics is ecology. *Bioscience* 44: 514.

Sample, V.A., N. Johnson, G.H. Aplet, and J.T. Olson. 1993. Defining sustainable forestry: introduction. In Aplet, G.H., N. Johnson, J.T. Olson, and V.A. Sample (eds). *Defining sustainable forestry*, pp. 3–10. Island Press, Washington, DC.

Schaef, A.W. 1992. *Beyond therapy, beyond science: a new model for healing the whole person.* Harper, San Francisco, CA.

Schaefer, M.B. 1956. The scientific basis for a conservation program. In *Papers presented at the International Technical Conference on the conservation of the living resources of the sea*, pp. 15–55 United Nations Publication 1956.II.B.1.

Schaeffer, D.J., E.E. Herricks, and H.W. Kerster. 1988. Ecosystem health: 1. Measuring ecosystem health. *Environmental Management* 12: 445–455.

Scheiner, S.M., A.J. Hudson, M.A. VanderMeulen. 1993. An epistemology for ecology. *Bulletin of the Eological Society of America* 74: 17–21.

Schindler, D.W. 1989. Biotic impoverishment at home and abroad. *Bioscience* 39: 426.

Schlesinger, W.H. 1989. Discussion: ecosystem structure and function. In Roughgarden, J., R.M. May, and S.A. Levin (eds). *Perspectives in ecological theory*, pp. 268–274. Princeton University Press, Princeton, NJ.

Schmid, P.E., M. Tokeshi, and J.M. Schmid-Araya. 2001. Relation between population density and body size in stream communities. *Science* 289: 1557–1560.

Schmidt-Nielsen, K. 1984. *Scaling: Why is animal size so important?* Cambridge University Press, New York, NY.

Schnute, J.T. and L.J. Richards. 2001. Use and abuse of fishery models. *Canadian Journal of Fisheries and Aquatic Sciences* 58: 10–17.

Schoener, T.W. 1985. Are lizard population sizes unusually constant through time? *American Naturalist* 126: 633–41.

Schoenly, K., R.A. Beaver, and T.A. Heumier. 1991. On the trophic relations of insects: a food-web approach. *American Naturalist* 137: 597–638.

Schultz, R.J. 1977. Evolution and ecology of unisex fishes. *Evolutionary Biology* 10: 277–331.

Scott, D.C. (ed.). 1912. Traditional history of the confederacy of the six nations. pp. 195–246. Section II, Proceedings and Transactions of the Royal Society of Canada, Vol. V, Series 3, Ottawa.

Seal, U.S. 1985. The realities of preserving species in captivity. In Hoage, R.J. (ed.). *Animal extinctions; what everyone should know*, pp. 71–95. Smithsonian Institution Press, Washington, DC.

Shapiro, D.Y. 1987. Differentiation and evolution of sex change in fishes. *Bioscience* 37: 490–497.

Sherman, K. 1994. Sustainability, biomass yields, and health of coastal ecosystems—an ecological perspective. *Marine Ecology-Progress Series* 112: 277–301.

Sibly, R.M., D. Barker, M.C. Denham, J. Hone, and M. Pagel. 2005. On the regualtion of populations of mammals, birds, fish and insects. *Science* 309: 607–610.

Signor, P.W. 1990. The geologic history of diversity. *Annual Review of Ecology and Systematics* 21: 509–539.

Silver, C.S. and R.S. DeFries (eds). 1990. *One earth, one future: Our changing global environment.* National Academy Press, Washington, DC.

Simberloff, D. 1980. A succession of paradigms in ecology: essentialism to materialism and probablism. *Synthese* 43: 3–39.

Simberloff, D. 1983. Sizes of coexisting species. In Futuyma, D.J. and M. Slatkin (eds). *Cooevolution*, pp. 404–430. Sinauer Associates, Sunderland, MA.

Simberloff, D. 1986. The proximate causes of extinction. In Raup, D.M. and D. Jablonski (eds). *Patterns and processes in the history of life*, pp. 259–276. Springer-Verlag, Berlin.

Simberloff, D. and W. Boecklen. 1991. Patterns of extinction in the introduced hawaiian avifauna: a reexamination of the role of competition. *American Naturalist* 138: 300–327.

Simpson, G.G. and W.S. Beck. 1965. *Life: An introduction to biology*. Harcourt, Brace and World, New York, NY.

Sinclair, A.R.E. 1996. Mammal populations: fluctuation, regulation, life history theory and their implications for conservation. In Floyd, R.B., A.W. Sheppard, and P.J. De Barro (eds), *Frontiers of population ecology*, pp. 127–154. CSIRO Publishing, Melbourne.

Slatkin, M. 1981. A diffusion model of species selection. *Paleobiology* 7: 421–425.

Slatkin, M. 1983. Genetic background. In Futuyma, D.J. and M. Slatkin (eds). *Coevolution*, pp. 14–32 Sinauer Associates, Sunderland, MA.

Smith, F.E. 1977. Comments revisited—or, what I wish I had said. In Innis, G.S. (ed.). *New directions in the analysis of ecological systems, Part 2*, pp. 231–235. The Society for Computer Simulation, La Jolla, CA.

Smith, H. 1958. *The religions of man*. Harper and Row, New York, NY.

Smith, H. 1994. *World religions: A guide to our wisdom traditions*. Harper, San Francisco, CA.

Smith, T.D. 1994. *Scaling in fisheries: The science of measuring the effects of fishing, 1855–1955*. Cambridge University Press, New York, NY.

Sobolevsky, Y.I. and O.A. Mathisen. 1996. Distribution, abundance, and trophic relationships of Bering Sea cetaceans. In Mathisen, O.A. and K.O. Coyle (eds). *Ecology of the Bering Sea: A review of Russian literature*, pp. 265–275. Alaska Sea Grant College Program Report No. 96–01. University of Alaska, Fairbanks.

Solan, M. B.J. Cardinale, A.L. Downing, K.A.M. Englehardt, J.L. Ruesink, and D.S. Srivastava. 2004. Extinction and ecosystem function in the marine benthos. *Science* 306: 1177–1180.

Solé, R.V. and J. Bascompte. 2006. *Self-organization in complex ecosystems*. Princeton University Press, Princeton, NJ.

Soulé, M.E. and B.A. Wilcox (eds). 1980. *Conservation biology, an evolutionary-ecological perspective*. Sinauer Associates, Sunderland, MA.

Southwood, T.R.E. 1977. Habitat, the templet for ecological strategies? *Journal of Animal Ecology* 46: 337–365.

Spellerberg, I.E. 1991. *Monitoring ecological change*. Cambridge University Press, Cambridge.

Spencer, P.D. and J.S. Collie. 1997. Patterns of population variability in marine fish stocks. *Fisheries Oceanography* 6: 188–204.

Spendolini, M.J. 1992. *The benchmarking book*. American Management Association, New York, NY.

Stanley, S.M. 1973. An Explanation for Cope's Rule. *Evolution* 27: 1–26.

Stanley, S.M. 1975a. A theory of evolution above the species level. *Proceedings of the National Academy of Sciences of the U.S.A.* 72: 646–650.

Stanley, S.M. 1975b. Clades versus clones in evolution: why we have sex. *Science* 190: 382–383.

Stanley, S.M. 1979. *Macroevolution, pattern and process*. W.H. Freeman and Co., San Francisco, CA.

Stanley, S.M. 1984. Marine mass extinctions: a dominant role for temperatures. In Nitecki, M.H. (ed.). *Extinctions*, pp. 69–118. The University of Chicago Press, Chicago, IL.

Stanley, S.M. 1985. Extinction as part of the natural evolutionary process: a paleobiological perspective. In Hoage, R.J. (ed.). *Animal extinctions; What everyone should know*, pp. 31–46. Smithsonian Institution Press, Washington, DC.

Stanley, S.M. 1989. Fossils, macroevolution and theoretical ecology. In Roughgarden, J., R.M. May, and S.A. Levin (eds). *Perspectives in ecological theory*, pp. 125–134. Princeton University Press, Princeton, NJ.

Stanley, S.M. 1990a. The species as a unit of large-scale evolution. In Warren, L. and H. Koprowski (eds). *New perspectives on evolution*, pp. 87–99. John Wiley and Sons, Inc., New York, NY.

Stanley, S.M. 1990b. The general correlation between rate of speciation and rate of extinction: fortuitous causal linkages. In Ross, R.M., and W.D. Allmon (eds). *Causes of evolution: a paleontological perspective*, pp. 103–127. University of Chicago Press, Chicago, IL.

Stanley, S.M., B. Van Valkenburgh, and R.S. Steneck. 1983. Coevolution and the fossil record. In Futuyma, D.J. and M. Slatkin (eds). *Cooevolution*, pp. 328–349. Sinauer Associates, Sunderland, MA.

Stanley, T.R., Jr. 1995. Ecosystem management and the arrogance of humanism. *Conservation Biology* 9: 255–262.

Stebbins, G.L. 1960. The comparative evolution of genetic systems. In Tax, S. (ed.). 1960. *Evolution after Darwin The evolution of life*, pp. 197–226. University of Chicago Press, Chicago, IL. 629pp.

Steffen, W., A. Sanderson, P.D. Tyson *et al.* 2004. *Global change and the earth system: a planet under pressure*. Springer-Verlag, Berlin.

Stenseth, N.C. 1985. Darwinian evolution in ecosystems—the Red Queen view. In Greenwood, P.J. P.H. Harvey, and M. Slatkin (eds). *Evolution: Essays in honor of John Maynard Smith*, pp. 55–72. Cambridge University Press, Cambridge.

Stenseth, N.C. 1986. Darwinian evolution in ecosystems: a survey of some ideas and difficulties together with some possible solutions. In Casti, J.L. and A. Karlqvist (eds). *Complexity, language and life: Mathematical approaches*, pp. 105–145. Springer-Verlag, Berlin.

Stenseth, N.C. 1989. On the evolutionary ecology of mammalian communities. In Morris, D.W., Z. Abramsky, B.J. Fox, and M.R. Willig (eds). *Patterns in the structure of mammalian communities*, pp. 219–228. Texas Tech University Press, Lubbock, TX.

Stenseth, N.C. and J. Maynard Smith. 1984. Coevolution in ecosystems; Red Queen evolution or stasis? *Evolution* 38: 870–880.

Stokes, K., R. Law. 2000. Fishing as an evolutionary force. *Marine Ecology Progress Series*. 208: 307–309.

Strickland, G.T. 1984. *Hunter's tropical medicine*. W.B. Saunders Co., Philadelphia, PA.

Strong, D.R., J.H. Lawton, and T.R.E. Southwood. 1984. *Insects on plants: Community patterns and mechanisms*. Blackwell Scientific, Oxford.

Sugihara, G., K. Schoenly, and A. Trombla. 1989. Scale invariance in food web properties. *Science* 245: 48–52.

Swain, D.P., A.F. Sinclair, and J.M. Hanson. 2007. Evolutionary response to size-selective mortality in an exploited fish population. *Proceedings of the Royal Society of London, Series B* 274: 1015–1022.

Swellengrebel, N.H. and M.M. Sterman. 1961. *Animal parasites in man*. Van Nostrand Co., Inc., NY.

Swimme, B. and T. Berry. 1994. *The universe story*. Harper, San Francisco, CA.

Tainter, J.A. 1988. *The collapse of complex societies*. Cambridge University Press, Cambridge.

Talbot, L.M. 2008. Introduction: the quest for environmental sustainability. In Rockford, L.L., R.E. Stewart, and T. Dietz (eds). *Foundations of environmental sustainability*, pp. 3–24. Oxford University Press, New York, NY.

Tamura, T. and S. Ohsumi. 1999. *Estimation of total food consumption by cetaceans in the world's oceans*. The Institute of Cetacean Research (ICR), Tokyo, Japan.

Taylor, L.R. and I.P. Woiwod. 1980. Temporal stability as a density-dependent species characteristic. *Journal of Animal Ecology* 49: 209–224.

Temple, S.A. 1977. Plant-animal mutualism: coevolution with dodo leads to near extinction of plant. *Science* 197: 885–886.

Terborgh, J. and C.P. Van Schaik. 1987. Convergence vs. non-convergence in primate communities. In Gee, J.H.R. and P.S. Giller (eds). *Organization of communities:*

Past and present, pp. 205–226. Blackwell Scientific Publications, Oxford.

Terborgh, J. and B. Winter. 1980. Some causes of extinction. In Soulé M.E. and M.A. Wilcox (eds). *Conservation biology, an evolutionary-ecological perspective*, pp. 119–133. Sinauer Associates, Sunderland, MA.

Thomas, C.D., A. Cameron, R.E. Green *et al.* 2004. Extinction risk from climate change. *Nature* 427: 145–148.

Thomas, W.L. (ed.). 1956. *Man's role in changing the face of the earth*. University of Chicago Press, Chicago, IL.

Thompson, J.N. 1982. *Interaction and coevolution*. John Wiley and Sons, New York, NY.

Thompson, J.N. 1994. *The coevolutionary process*. University of Chicago Press, Chicago, IL.

Thompson, J.N. 2005. *The geographic mosaic of coevolution*. University of Chicago Press, Chicago, IL.

Thorne-Miller, B. and J. Catena. 1991. *The living ocean, understanding and protecting marine biodiversity*. Island Press, Washington, DC.

Tiger, L. and R. Fox. 1989. *The imperial animal*. Holt, Rinehart and Winston, New York, NY.

Tilman, D., R.M. May, C.L. Lehman, and M.A. Nowak. 1994. Habitat destruction and the extinction debt. *Nature* 371: 65–66.

Towell, R.G., R.R. Ream, and A.E. York. 2006. Decline in northern fur seal (*Callorhinus ursinus*) pup production on the Pribilof Islands. *Marine Mammal Science* 22: 486–491.

Travis, J. and L.D. Mueller. 1989. Blending ecology and genetics: progress toward a unified population biology. In Roughgarden, J., R.M. May, and S.A. Levin (eds). *Perspectives in ecological theory*, pp. 101–124. Princeton University Press, Princeton, NJ.

Trotter, M.M. and B. McCulloch. 1984. Moas, men and middens. In Martin, P.S. and R.G. Klein (eds). *Quaternary extinctions: a prehistoric revolution*, pp. 708–727. University of Arizona Press, Tucson, AZ.

Tudge, C. 1989. The rise and fall of *Homo sapiens sapiens*. *Philosophical Transactions of the Royal Society of London, Series B* 325: 479–488.

Turner, B.L., II, W.C. Clark, R.W. Kates, J.F. Richards, J.T. Mathews, and W.B. Meyer (eds). 1990. *The earth as transformed by human action; global and regional changes in the biosphere over the past 300 years*. Cambridge University Press, New York, NY.

Ueckert, D.N. and R.M. Hansen. 1971. Dietary overlap of grasshoppers on sandhill rangeland in northeastern Colorado. *Oecologia* 8: 276–295.

Ulanowicz, R.E. 1989. Energy flow and productivity in the oceans. In Grubb, P.J. and J.B. Whittaker (eds). *Toward a more exact ecology*, pp. 327–351. Blackwell Scientific, Oxford, UK.

Ulanowicz, R.E. 1992. Ecosystem health and trophic flow networks. In Costanza, R., B.G. Norton, and B.D. Haskell (eds). *Ecosystem health: New goals for environmental management*, pp. 190–206. Island Press, Washington, DC.

Union of Concerned Scientists. 1992. *World scientists' warning to humanity.* Available UCS Headquarters, 26 Church Street, Cambridge, MA 02238.

Valkenburgh, B.V., X. Wang, and J. Damuth. 2004. Cope's rule, hypercarnivory, and extinction in North American canids. *Science* 306: 101–104.

Vallentyne, J.R. 1993. Biospheric foundations of the ecosystem approach to environmental management. *Journal of Aquatic Ecosystem Health* 2: 9–13.

van der Maarel, E. 1975. Man-made natural ecosystems in environmental management and planning. In van Dobben, W.H. and R.H. Lowe-McConnell (eds). *Unifying concepts in ecology*, pp. 263–274. Dr. W. Junk b. v., The Hague, the Netherlands, and the Centre for Agricultural Publishing and Documentation, Wageningen, the Netherlands.

van Dobben, W.H. and R.H Lowe-McConnell. (eds.) 1975. Preface. *Unifying concepts in ecology*, pp. 5–8. Dr. W. Junk b. v., The Hague, the Netherlands, and the Centre for Agricultural Publishing and Documentation, Wageningen, the Netherlands.

Van Nostrand's Scientific Encyclopedia. 1958. *Explosives.* Van Nostrand, Princeton, NJ.

Van Valen, L. 1971. Group selection and the evolution of dispersal. *Evolution* 25: 591–598.

Van Valen, L. 1973a. Body size and numbers of plants and animals. *Evolution* 27: 27–35.

Van Valen, L. 1973b. A new evolutionary law. *Evolutionary Theory* 1: 1–30.

Vermeij, G.J. 1983. Intimate associations and coevolution in the sea. In Futuyma, D.G. and M. Slatkin (eds). *Coevolution*, pp. 311–327. Sinauer Associates Sunderland, MA.

Vermeij, G.J. 1987. The dispersal barrier in the tropical Pacific: implications for molluscan speciation and extinction. *Evolution* 41: 1046–1058.

Vitousek, P.M. 1994. Beyond global warming: ecology and global change. *Ecology* 75: 1861–1876.

Vitousek, P.M., P.R. Ehrlich, A.H. Ehrlich, and P.A. Matson. 1986. Human appropriation of the products of photosynthesis. *Bioscience* 36: 368–373.

Vitousek, P.M., W.A. Reiners, J.M. Melillo, C.C. Grier, and J.R. Gosz. 1981. Nitrogen cycling and loss following forest perturbation: the components of response. In Barrett, G.W. and R. Rosenberg (eds). *Stress effects on natural ecosystems*, pp. 115–128. John Wiley and Sons, New York, NY.

Vitousek P.M., J.D. Aber, R.W. Howarth, G.E. Likens, P.A. Matson, D.W. Schindler, W.H. Schlesinger, and D.G. Tilman. 1997. Human alteration of the global nitrogen cycle: Sources and consequences. *Ecological Applications* 7: 737–750.

Vörörsmarty, C.J., P. Green, J. Salisbury, and R.B. Lammers. 2000. Global water resources: vulnerability from climate change and population growth. *Science* 289: 284–288.

Vrba, E.S. 1980. Evolution, species and fossils: how does life evolve? *South African Journal of Science* 76: 61–84

Vrba, E.S. 1983. Macroevolutionary trends: new perspectives on the roles of adaptation and incidental effect. *Science* 221: 387–389.

Vrba, E.S. 1984. What is species selection? *Systematic Zoology* 33: 318–328.

Vrba, E.S. 1989. Levels of selection and sorting with special reference to the species level. *Oxford Surveys in Evolutionary Biology* 6: 111–168.

Vrba, E.S. and N. Eldredge. 1984. Individuals, processes and hierarchy: towards a more complete evolutionary theory. *Paleobiology* 10: 146–171.

Vrba, E.S. and S.J. Gould. 1986. The hierarchical expansion of sorting and selection: sorting and selection cannot be equated. *Paleobiology* 12: 217–228.

Wagner, F.H. 1977. Species vs. ecosystem management. Transactions of the 42nd North *American Wildlife and Natural Resources Conference* 42: 14–42.

Waldrop, M.M. 1992. *Complexity: The emerging science at the edge of order and chaos.* Simon and Schuster, New York, NY.

Walker, A. 1984. Extinction in Hominid evolution. In Nitecki, M.H. (ed.). *Extinctions*, pp. 119–152. The University of Chicago Press, Chicago, IL.

Wallace, P.A.W. 1946. *The white roots of peace.* University of Pennsylvania Press, Philadelphia, PA.

Wallace, R.L. 1994. *The marine mammal commission compendium of selected treaties, international agreements, and other relevant documents on marine resources, wildlife, and the environment.* US Marine Mammal Commission, Washington, DC (Three Vols).

Walters, C. 1986. *Adaptive management of renewable resources.* Macmillan, New York, NY.

Walters, C. 1992. Perspectives on adaptive policy design in fisheries management. In Jain, S.K and L.W. Botsford (eds). *Applied population biology*, pp. 249–262. Kluwer Academic, Dordrecht, Netherlands.

Warburton, K. 1989. Ecological and phylogenetic constraints on body size in Indo-Pacific fishes. *Environmental Biology of Fishes* 24: 13–22.

Ward, P.D. 2007. *Under a green sky: Global warming, the mass extinctions of the past, and what they mean for our future.* Smithsonian Books/Collins, New York, NY.

Watts, A.W. 1951. *The wisdom of insecurity: A message for an age of anxiety.* Vintage Books, New York, NY.

Werbos, P.J. 1992. Energy and population: transitional issues and eventual limits. In Grant, L. (ed.). *Elephants in the Volkswagon,* pp. 32–49. W.H. Freeman and Co., San Francisco, CA.

Werren, J.H. 1987. Labile sex ratios in wasps and bees. *Bioscience* 37: 498–506.

Western, D. and M.C. Pearl (eds). 1989. *Conservation for the twenty-first century.* Oxford University Press, New York, NY.

Western, D. and R.M. Wright (eds). 1994. *Natural connections: perspectives in community-based conservation.* Island Press, Washington, DC.

Westman, W.E. 1990. Detecting early signs of regional air-pollution injury to coastal sage scrub. In Woodwell, G.M. (ed.). *The earth in transition; patterns and processes of biotic impoverishment,* pp. 323–346. Cambridge University Press, New York, NY.

Whelpton, P.K. 1939. Population policy for the United States. *Journal of Heredity* 30: 401–406.

White, L., Jr. 1967. The historical roots of our ecological crisis. *Science* 155: 1203–1207.

White, M.J.D. 1978. *Modes of speciation.* Freeman and Co., San Francisco, CA.

Whitehouse, P.L., M.B. Allen, and G.A. Milne. 2007. Glacial isostatic adjustment as a control on coastal processes: an example from the Siberian Arctic. *Geology* 35: 747–750.

Whitfield, C.L. 1987. *Healing the child within: discovery and recovery for adult children of dysfunctional families.* Health Communications, Inc., Pompano Beach, FL.

Whitmore, T.C. 1980. The conservation of tropical rain forest. In Soulé, M.E. and M.A. Wilcox (eds). *Conservation biology, an evolutionary-ecological perspective,* pp. 303–318. Sinauer Associates, Sunderland, MA.

Whittaker, R.H. 1975. *Communities and ecosystems* (2nd ed.). McMillan Publishing Co., New York, NY.

WHO Expert Committee on Bacterial, and Viral Zoonoses. 1982. *Bacterial and viral zoonoses.* WHO Technical report no. 682. WHO, Geneva.

WHO Scientific Group on Arthropod-born, and rodent-borne viral diseases. 1985. *Arthropod-borne and rodent-borne viral diseases.* WHO Scientific Group report no. 719. WHO, Geneva.

WHO Scientific Group on the Biology of Malaria Parasites. 1987. *The biology of malaria parasites.* WHO Technical Report Series, report no. 743. WHO, Geneva.

Wignall, P.B. 2001. Large igneous provinces and mass extinctions. *Earth-Science Reviews* 53: 1–33.

Wilber, K. 1995. *Sex, ecology, spirituality: the spirit of evolution.* Shambhala Publications, Boston, MA.

Wilber, K. 1996. *A brief history of everything.* Shambala, Boston.

Wilcox, B.A. and D.D. Murphy. 1985. Conservation strategy: the effects of fragmentation on extinction. *American Naturalist* 125: 879–887.

Williams, G.C. 1975. *Sex and evolution.* Princeton University Press, Princeton, NJ.

Williams, G.C. 1992. *Natural selection: Domains, levels, and challenges.* Oxford University Press, New York, NY.

Willis, D. 1995. *The sand dollar and the slide rule: Drawing blueprints from nature.* Addison-Wesley, Reading, MA.

Wilson, D.S. 1980. *The natural selection of populations and communities.* Benjamin/Cummings, Menlo Park, CA.

Wilson, D.S. 1983. The group selection controversy: history and current status. *Annual Review of Ecology and Systematics* 14: 159–188.

Wilson, E.O. 1975. The origin of sex. *Science* 188: 139–140.

Wilson, E.O. 1985. The biological diversity crisis: a challenge to science. *Issues in Science and Technology* 2: 20–29.

Wilson, E.O. 1998. *Consilience: The unity of knowledge.* Knopf, New York. 332pp.

Wilson, J.W. and O.A. Plunlatt. 1965. *The fungus diseases of man.* University of California Press, Berkeley, CA.

Wilson, M.E. 1991. *A world guide to infections: Diseases, distribution and diagnosis.* Oxford University Press, New York, NY.

Wood, C.A. 1994. Ecosystem management: achieving the new land ethic. *Renewable Resources Journal* 12: 6–12.

Woodley, S., J.J. Kay and G.R. Francis (eds). 1993. *Ecological integrity and the management of ecosystems.* St. Luicie press, Delray Beach, FL.

Woodwell, G.M. 1970. Effects of pollution on the structure and physiology of ecosystems. *Science* 168: 429–433.

Woodwell, G.M. 1990. The earth under stress: a transition to climatic instability raises questions about patterns of impoverishment. In Woodwell, G.M. (ed.). *The earth in transition; patterns and processes of biotic impoverishment,* pp. 3–8. Cambridge University Press, New York, NY.

Woodwell, G.M. and R.A. Houghton. 1990. The experimental impoverishment of natural communities: effects of ionizing radiation on plant communities, 1961–1976. In Woodwell, G.M. (ed.). *The earth in transition; patterns and processes of biotic impoverishment,* pp. 9–24. Cambridge University Press, New York, NY.

World Almanac. 1993. *The world almanac and book of world facts, 1994.* Funk and Wagnalls, Mahwah, NJ.

World Conservation Monitoring Centre. 1992. *Global biodiversity.* Chapman & Hall, New York, NY.

Wright, D.H. 1990. Human impacts on energy flow through natural ecosystems, and implications for species endangerment. *Ambio* 19: 189–194.

Wright, S. 1956. Modes of selection. *American Naturalist* 90: 5–24.

Wright, S. 1967. Comments on the preliminary working papers of Eden and Waddington. *Wistar Institute symposium monograph* 5: 117–120.

Wynne-Edwards, V.C. 1962. *Animal dispersion in relation to social behaviour.* Hafner Publishing, New York, NY.

Yan, N.D. and P.M., Welbourn. 1990. The impoverishment of aquatic communities by smelter activities near Sudbury, Canada. In Woodwell, G.M. (ed.). *The earth in transition; patterns and processes of biotic impoverishment*, pp. 477–494. Cambridge University Press, New York, NY.

Yodzis, P. 1980. The connectance of real ecosystems. *Nature* 284: 544–545.

Yodzis, P. 1984. Energy flow and the vertical structure of real ecosystems. *Oecologia* 65: 86–88.

Yoffee, N. and G.L. Cowgill (eds). 1988. *The collapse of ancient states and civilizations.* University of Arizona Press, Tucson, AZ.

Zahavi, A. and A. Zahavi. 1999. *The handicap principle: A missing piece of Darwin's puzzle.* Oxford University Press, New York, NY.

Zeveloff, S.I. and M.S. Boyce. 1988. Body size patterns in North American mammal faunas. In Boyce, M.S. (ed.). 1988. *Evolution of life histories of mammals*, pp. 123–146. Yale University Press, New Haven, Conn.

Zimberoff, D. and D. Hartman. 2001. Four primary existential themes. *Journal of Heart-centered Therapies* 4: 3–64.

Zuckerman, A.J., J.E. Banatvala, and J.R. Pattison (eds). 1990. *Principles and practice of clinical virology.* John Wiley & Sons, New York, NY.

Author Index

Figures and tables are indexed in bold.